It is usually straightforward to calculate the result of a practical experiment in the laboratory. Estimating the accuracy of that result is often regarded by students as an obscure and tedious routine, involving much arithmetic. An estimate of the error is, however, an integral part of the presentation of the results of experiments.

This textbook is intended for undergraduates who are performing laboratory experiments in the physical sciences and who have to calculate errors for the first time. It is not a formal textbook on statistics, but is a practical guide on how to analyse data and estimate errors. The necessary formulae for performing calculations are given, and the ideas behind them are explained. Specific examples are worked through step by step in the text. Emphasis is placed on the need to think about whether a calculated error is sensible.

At first students should take this book with them to the laboratory, and the format is intended to make this convenient. The book will provide the necessary understanding of what is involved, should inspire confidence in the method of estimating errors, and enable numerical calculations to be performed without too much effort. The author's aim is to make practical classes more enjoyable. Students who use this book will be able to complete their calculations quickly and confidently, leaving time to appreciate the basic physical ideas involved in the experiments.

A practical guide to data analysis for physical science students

A practical guide to data analysis for physical science students

LOUIS LYONS

Department of Physics
University of Oxford

Published by the Press Syndicate of the University of Cambridge
The Pitt Building, Trumpington Street, Cambridge CB2 1RP
40 West 20th Street, New York, NY 10011-4211, USA
10 Stamford Road, Oakleigh, Victoria 3166, Australia

First published 1991
Reprinted 1992

Printed in Great Britain at the University Printing House

A catalogue record of this book is available at the British Library

Library of Congress cataloguing in publication data
Lyons, Louis.
A practical guide to data analysis for physical
science students/ Louis Lyons.
p. cm.
ISBN 0-521-41415-6. – ISBN 0-521-42463-1 (pbk.)
1. Physics–Experiments–Statistical methods. I. Title.
QC33.L9 1991
500.2′0724–dc20 91-1962 CIP

ISBN 0 521 41415 6 hardback
ISBN 0 521 42463 1 paperback

שְׁגִיאוֹת מִי יָבִין

Who can comprehend errors?

Psalms 19:13

Contents

Preface

This short book is intended to be a practical guide, providing sets of rules that will help you to analyse the data you collect in your regular experimental sessions in the laboratory. Even more important, explanations and examples are provided to help you understand the ideas behind the formulae. Emphasis is also placed on thinking about the answers that you obtain, and on helping you get a feeling for whether they are sensible.

In contrast, this does not set out to be a text on statistics, and certainly not to be a complete course on the subject. Also, no attempt is made to provide rigorous mathematical proofs of many of the required formulae. These are important, and if required can be consulted in any standard textbook on the subject.

I believe that it will be necessary to read this material more than once. You really need to have understood the ideas involved before you do your first practical; but on the other hand, it would be much easier to absorb the material after you have actually done a couple of experiments and grappled with problems of trying to do the analysis yourself. Thus it is a good idea to read the book quickly, so that you at least discover what topics are covered and where to find them again when you need them. At this stage, you need not worry if not everything is entirely comprehensible. Then you take the book with you into your practicals, so that you can refer to it for help with each of your early calculations. As you become experienced, you will need to consult it less and less. However, it is a good idea to return to reading the whole book, this time aiming to understand it almost completely.

Since data analysis is a very practical subject, the only way to become proficient is to be involved in actually doing it. To help you achieve this, there are a few problems at the end of each chapter. You are strongly recommended to solve them. Omitting them is analogous to trying to learn to swim by reading a book on the subject without ever getting into the water.

This book is a slightly extended version of a short series of lectures I have been giving to first year Oxford Physics students. The material for these lectures was based on experience I have gained over several years by supervising and marking the weekly practical work of undergraduates here. It is clear that calculating the accuracy of an experimental result often becomes the most difficult part of the assignment. Many students regard the necessary calculations as obscure and complicated. The aim of this book is to dispel these ideas, to make it clear what the correct procedure is, to help you avoid excessively long calculations and to enable you to realise when your calculations have yielded ridiculous answers.

I hope that this book will be a useful companion that will assist you with your error calculations and data analysis, and to stop them being a chore. In this way, you should be able to enjoy your practical work, and to devote your energy to understanding the basic physical ideas involved.

I am grateful to the Numerical Algorithms Group (NAG) for permission to use routines from their program library for producing the Tables in Appendices 6 and 7.

Louis Lyons
Oxford, 1991

Glossary and Conventions

μ = true mean of a distribution (or of a large population)
$\bar{x}$ = estimated mean for a sample
σ^2 = true variance of a distribution (or of a large population)
s^2 = variance as estimated from the spread of observations in a sample
u^2 = variance of the estimated mean

Experimental results are usually quoted in the form $y \pm \varepsilon$. The estimate y is said to be unbiassed provided that, were we to repeat our experiment N times, the average of the y values for the whole set of measurements would tend to the true (but generally unknown) value y_0 as N becomes larger and larger. The quantity ε is our estimate of the standard deviation (RMS deviation from the mean) of the distribution of results that we would expect if we repeated the experiment with similar apparatus many times. We refer to ε as 'the error' on the result. Occasionally results are given in the form $y \pm \varepsilon \pm \delta$. Here ε is the contribution from the random (or statistical) error, and δ that from systematic effects.

For a result like 20 ± 1 cm, we refer to the 1 cm as the 'absolute error', while the 'fractional error' is $1/20 = 5\%$.

The average of y is denoted by $\bar{y}$ or $\langle y \rangle$.

The sign $\sum$ means that we have to perform a summation. Thus

$$\sum_{i=1}^{N} y_i = y_1 + y_2 + \cdots + y_N.$$

Then the average $\bar{y} = \sum y_i / N$.

1

Experimental errors

1.1 Why estimate errors?

When performing experiments at school, we usually considered that the job was over once we obtained a numerical value for the quantity we were trying to measure. At university, and even more so in everyday situations in the laboratory, we are concerned not only with the answer but also with its accuracy. This accuracy is expressed by quoting an experimental error on the quantity of interest. Thus a determination of the acceleration due to gravity in our laboratory might yield an answer

$$g = (9.70 \pm 0.15)\ \mathrm{m/s^2}.$$

In Section 1.4, we will say more specifically what we mean by the error of ± 0.15. At this stage it is sufficient to state that the more accurate the experiment the smaller the error; and that the numerical value of the error gives an indication of how far from the true answer this particular experiment may be.

The reason we are so insistent on every measurement including an error estimate is as follows. Scientists are rarely interested in measurement for its own sake, but more often will use it to test a theory, to compare with other experiments measuring the same quantity, to use this parameter to help predict the result of a different experiment, and so on. Then the numerical value of the error becomes crucial in the interpretation of the result.

For example, maybe we measured the acceleration due to gravity in order to compare it with the value of 9.81 m/s^2,* measured in another laboratory a few miles away last year. We could be doing this in order to see whether there had been some dramatic change in the gravitational constant G over the intervening period; to try to detect a large gold mine which could affect the gravitational field in our neighbourhood; to find out if the earth had stopped spinning (although there are easier ways of doing this); to discover the existence of a new force in nature which could make the period of a pendulum depend on the local topography, etc.

With a measurement of 9.70 m/s^2, do we have evidence for a discrepancy? There are essentially three possibilities.

Possibility 1
If as suggested above the experimental error is ± 0.15, then our determination looks satisfactorily in agreement with the expected value,

i.e. 9.70 ± 0.15 is consistent with 9.81.

Possibility 2
If we had performed a much more accurate experiment and had succeeded in reducing the experimental error to ± 0.01, then our measurement is inconsistent with the previous value. Hence, we should worry whether our experimental result and/or the error estimate are wrong. Alternatively, we may have made a world shattering discovery.

i.e. 9.70 ± 0.01 is inconsistent with 9.81.

Possibility 3
If we had been stupid enough to time only one swing of the pendulum, then the error on g could have been as large as ± 5. Our result is now consistent with expectation, but the accuracy is so low that it would be incapable of detecting even quite significant differences.

i.e. 9.70 ± 5 is consistent with 9.81,
and with many other values too.

Thus for a given result of our experiment, our reaction – 'Our measurement is in good shape' OR 'We have made a world shattering discovery' OR 'We should find out how to do better experiments' – depends on the

* Since this is an experimental number, it too has an uncertainty, but we assume that it has been measured so well that we can effectively forget about it here.

numerical estimate of the accuracy of our experiment. Conversely, if we know only that the result of the experiment is that the value of g was determined as 9.70 m/s^2 (but do not know the value of the experimental error), then we are completely unable to judge the significance of this result.

The moral is clear. Whenever you determine a parameter, estimate the error or your experiment is useless.

A similar remark applies to 'null measurements'. These occur in situations where you investigate whether changing the conditions of an experiment affects its result. For example, if you increase the amplitude of swing of your pendulum, does the period change? If, to the accuracy with which you can make measurements, you see no effect, it is tempting to record that 'No change was seen'. However this in itself is not a helpful statement. It may become important at some later stage to know whether the period was constant to within 1%, or perhaps within 1 part in a million. Thus, for example, the period is expected to depend slightly on the amplitude of swing, and we may be interested to know whether our observations are consistent with the expected change. Alternatively we may need to know how accurate the pendulum is as a clock, given that its amplitude is sometimes 10° and at others 5°. With simply the statement 'No change was seen', we have no idea at all of what magnitude of variation of the period could be ruled out. It is thus essential in these situations to give an idea of the maximum change that we would have been capable of detecting. This could consist of a statement like 'No change was observed; the maximum possible change in period was less than 1 part in 300'.

It is worth remembering that null measurements, with sufficiently good limits on the possible change, have sometimes led to real progress. Thus, at the end of the last century, Michelson and Morley performed an experiment to measure the speed of the earth through the hypothesised aether. This would have produced shifts in the optical interference fringe pattern produced in their apparatus. They observed no such shift, and the limit they were able to place on the effect was sufficiently stringent that the idea of the aether was discarded. The absence of an aether was one of the cornerstones on which Einstein's Special Theory of Relativity was built.

Thus 'null observations' can be far from useless, provided you specify what the maximum possible value of the effect could have been.

1.2 Random and systematic errors

1.2.1 What they are

There are two fundamentally different sorts of errors associated with any measurement procedure, namely random (or statistical) and systematic errors. Random errors come from the inability of any measuring device (and the scientist using it) to give infinitely accurate answers.* Another source of random errors is the fluctuations that occur in observations on a small sample drawn from a large population. On the other hand, systematic errors result in measurements that for one reason or another are simply wrong. Thus when we make a series of repeated measurements, the effect of random errors is to produce a spread of answers scattered around the true value. In contrast, systematic errors can cause the measurements to be offset from the correct value, even though the individual results can be consistent with each other. (See Fig. 1.1.)

Thus, for example, suppose someone asks you the exact time. You look at your watch, which has only hour and minute hands, but no second hand. So when you try to estimate the time, you will have a random error of something of the order of a minute. You certainly would have extreme difficulty trying to be precise to the nearest second. In addition to this random error, there may well be systematic errors too. For example, your watch may be running slow, so that it is wrong by an amount that you are not aware of but may in fact be 10 minutes. Again, you may recently have come back home to England from Switzerland, and forgotten to reset your watch, so that it is out by 1 hour. As is apparent from this example, the random error is easier to estimate, but there is the danger that if you are not careful you may be completely unaware of the more important systematic effects.

As a more laboratory oriented example, we now consider an experiment designed to measure the value of an unknown resistor, whose resistance R_2 is determined as

$$R_2 = \frac{V_2 - V_1}{V_1} R_1 \tag{1.1}$$

(see Fig. 1.2). Thus we have to measure the voltages V_1 and V_2, and the

* Except possibly for the situation where we are measuring something that is integral (e.g., the number of cosmic rays passing through a small detector during one minute). See, however, the next sentence of the text, and the remarks about Poisson distributions in Section 1.2.2.

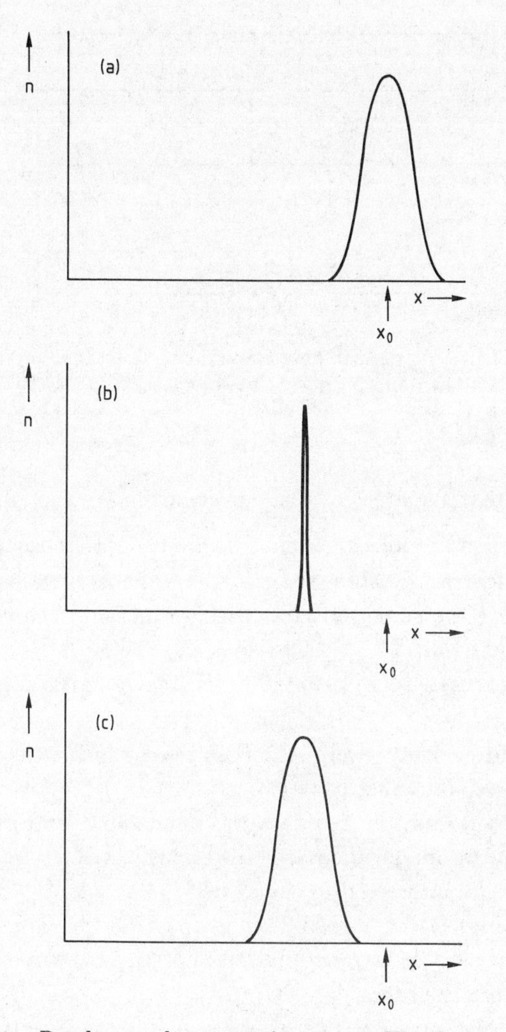

Fig. 1.1. Random and systematic errors. The figures show the results of repeated measurements of some quantity x whose true value is shown by the arrows. The effect of random errors is to produce a spread of measurements, centred on x_o (see (a)). On the other hand, systematic effects (b) can shift the results, while not necessarily producing a spread. Finally, the effect of random and systematic errors, shown in (c), is to produce a distribution of answers, centred away from x_o.

other resistance R_1. The random errors are those associated with the measurements of these quantities.

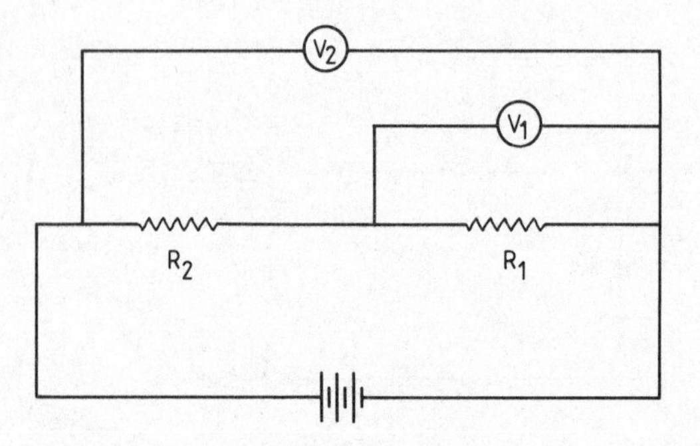

Fig. 1.2. Circuit for determining an unknown resistance R_2 in terms of a known one R_1 and the two voltages V_1 and V_2.

The most obvious sources of systematic errors are the following.

(i) The meters or oscilloscopes that we are using to measure V_1 and V_2 may be incorrectly calibrated. How this affects the answer depends on whether the same device is used to measure the two voltages. (See section 1.8.)

(ii) The meter used to measure the resistor R_1 may similarly be in error.

(iii) If our voltage source were AC, then stray capacitances and/or inductances could affect our answer.

(iv) The resistors may be temperature dependent, and our measurement may be made under conditions which differ from those for which we are interested in the answer.

(v) The impedances of the voltmeters may not be large enough for the validity of the approximation that the currents through the resistors are the same.

(vi) Electrical pick-up could affect the readings of the voltmeters.

Systematic errors can thus arise on any of the actual measurements that are required in order to calculate the final answer (e.g. points (i) and (ii) above). Alternatively, they can be due to more indirect causes; thus effects (iii)–(vi) are produced not by our instruments being incorrect, but more by the fact that we are not measuring exactly what we are supposed to.

In other situations it might be that there are implicit assumptions in the derivation of the equation on which we are relying for obtaining our answer. For example, the period of a pendulum of length l is $2\pi\sqrt{l/g}$ only if the amplitude of oscillations is small, if we can neglect air resistance, if the top of the pendulum is rigidly secured, etc. If these are not true, then our experiment has a systematic error. Whether such effects are significant or not depends on their magnitude compared with those of the random errors.

1.2.2 Estimating random errors

A recurring theme in this book is the necessity of providing error estimates on any measured quantity. Because of their nature, random errors will make themselves apparent by producing somewhat different values of the measured parameter in a series of repeated measurements. The estimated accuracy of the parameter can then be obtained from the spread in measured values as described in Section 1.4.

An alternative method of estimating the accuracy of the answer exists in cases where the spread of measurements arises because of the limited accuracy of measuring devices. The estimates of the uncertainties of such individual measurements can be combined as explained in Section 1.7 in order to derive the uncertainty of the final calculated parameter. This approach can be used in situations where a repeated set of measurements is not available for the method described in the previous paragraph. In cases where both approaches can be used, they should of course yield consistent answers.

The accuracy of our measurements will in general play little part in determining the accuracy of the final parameter in those situations in which the measurements are made on a population which exhibits its own natural spread of values. For example, the heights of ten-year-old children are scattered by an amount which is larger than the uncertainty with which the height of any single child can be measured. It is then this scatter and the sample size which determine the accuracy of the answer.

A similar situation arises where the observation consists in counting independent random events in a given interval. The spread of values will usually be larger than the accuracy of counting (which may well be exact); for an expected number of observations n, the spread is $\sqrt{n}$. This can be derived from the properties of the Poisson distribution, which is discussed in Appendix 4.

Another example is provided by the measurement of the mean lifetime τ of a radioactive element. This we can do by finding the average of the observed decay times of a sample of the disintegrations. The nature of radioactivity is such that not all decays occur at the identical time τ, but in fact a large number would follow an exponential distribution (see

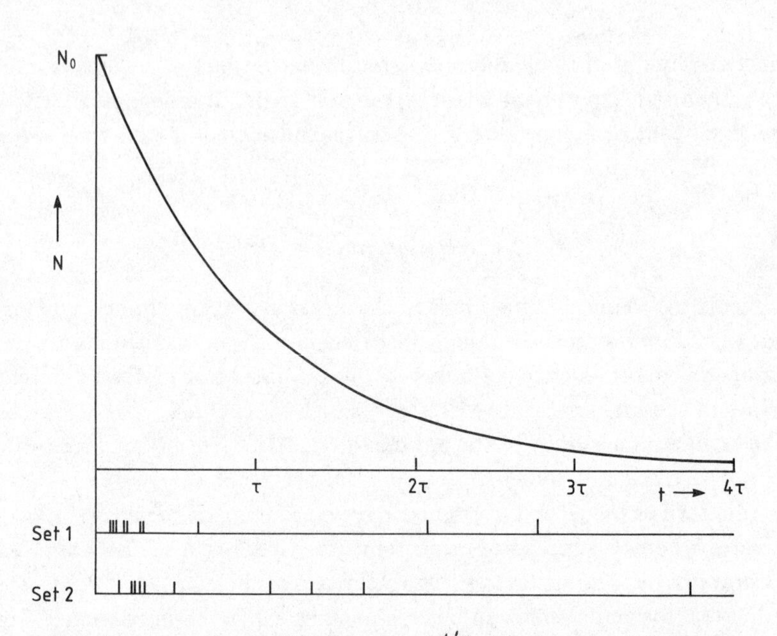

Fig. 1.3. An exponential $N = N_o\ e^{-t/\tau}$, for the expected distribution of decay times t of radioactive disintegrations of a source of mean lifetime τ. The bars below the t axis give two possible sets of observed decay times in experiments where each detected ten decays. The means of these times for the two samples are 0.68τ and 0.96τ. They differ from τ because of the statistical fluctuations associated with small samples.

Fig. 1.3). Thus the observed times for a small number of decays could fluctuate significantly if we repeated the experiment. This variation is a random effect, and is not connected with the accuracy with which we can measure individual decay times, which could be very much better than τ.

1.2.3 Worrying about systematic errors

For systematic errors, the 'repeated measurement' approach will not work; if our ohmeter is reading in kilohms while we think it is in ohms, the resistance will come out too small by a factor of ~1000 each time we repeat the experiment, and yet everything will look consistent.

Ideally, of course, all systematic effects should be absent. But if it is thought that such a distortion may be present, then at least some attempt can be made to estimate its importance and to correct for it. Thus if we suspect a systematic error on the ohmeter, we can check it by measuring some known resistors. Alternatively, if we are worried that the amplitude of our pendulum is too large, we can measure the period for different initial displacements, and then extrapolate our answer to the limit of a very small amplitude. In effect, we are then converting what was previously a systematic error into what is hopefully only a random one.

One possible check that can sometimes be helpful is to use constraints that may be relevant to the particular problem. For example, we may want to know whether a certain protractor has been correctly calibrated. One possible test is to use this protractor to measure the sum of the angles of a triangle. If our answer differs significantly from 180°, our protractor may be in error.

In general, there are no simple rules or prescriptions for eliminating systematic errors. To a large extent it requires common sense plus experience to know what are the possible dangerous sources of errors of this type.

Random errors are usually more amenable to methodical study, and the rest of this chapter is largely devoted to them. Nevertheless, it is important to remember that in many situations the accuracy of a measurement is dominated by the possible systematic error of the instrument, rather than by the precision with which you can actually make the reading.

Finally we assert that a good experimentalist is one who minimises and realistically estimates the random errors of his apparatus, while reducing the effect of systematic errors to a much smaller level.

1.3 Distributions

In Section 1.6 we are going to consider in more detail what is meant by the error σ on a measurement. However, since this is related to the concept of the spread of values obtained from a set of repeated measurements, whose distribution will often resemble a Gaussian (or normal) distribution, we will first have three mathematical digressions into the subjects of (a) distributions in general, (b) the mean and variance of a distribution, and (c) the Gaussian distribution.

A distribution $n(x)$ will describe how often a value of the variable x occurs in a defined sample. The variable x could be continuous or discrete, and its values could be confined to a finite range (e.g. 0–1) or

Table 1.1. *Examples of distributions*

Character	**Limits**	**x variable**	$n(x)$
	$1 \to \infty$	Integer x	No. of times you have produced a completely debugged computer program after x compilations
Discrete	$1 \to 7$	Day of week	No. of marriages on day x
	-13.6 eV $\to 0$	Energies of ground and excited states of hydrogen atoms	No. of atoms with electrons in state of energy x in atomic hydrogen at 30000°
	$-\infty \to \infty$	Measured value of parameter	No. of times measurment x is observed
Continuous	$0 \to \infty$	Time it takes to solve all problems in this book	No. of readers taking time x
	$0 \to 24$ hours	Hours sleep each night	No. of people sleeping for time x

could extend to $\pm\infty$ (or could occupy a semi-infinite range, e.g. positive values only). Some examples are given in Table 1.1.

As an example, Fig. 1.4 shows possible distributions of a continuous variable, the height h of 30-year-old men. If only a few values are available, the data can be presented by marking a bar along the h axis for each measurement (see Fig. 1.4(a)). In Fig. 1.4(b), the same data is shown as a histogram, where a fairly wide bin size for h is used and the vertical axis is labelled as n, the number of observations per centimetre interval of h, despite the fact that the bin size Δh used is 10 cm. The actual number of men corresponding to a given bin is $n\Delta h$, and the total number of men appearing in the histogram is $\sum n\Delta h$. If 100 times more measurements were available, the number of entries in each bin of the histogram would increase by a large factor (Fig. 1.4(c)), but it would now become sensible to draw the histogram with smaller bins, in order to display the shape of the distribution with better resolution. Because

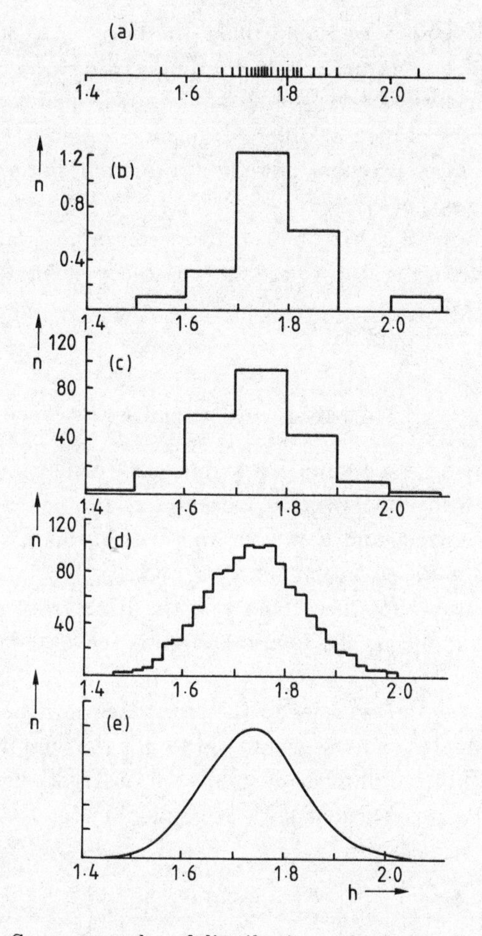

Fig. 1.4. Some examples of distributions of, say, the heights h (in metres) of 30-year-old men. (a) With only a few observations, each one is represented as a bar at the relevant position along the h axis. (b) The data of (a) could alternatively be drawn as a histogram, where n is the number of men per centimetre interval of h, even though the bin size in h is 10 cm. (c) A histogram as in (b), but with 100 times more observations. (d) The same data as in (c), but drawn with smaller bins of h. n is still the number of men per centimetre interval of h. (e) For an even larger number of observations and with smaller bin size, the histogram of (d) approaches a continuous distribution.

we plot $n(h)$ as the number of observations per centimetre, irrespective of bin size, the overall height of the histogram does not change much when we change the bin size (see Fig. 1.4(d)). Finally, for an even larger number of observations, we could make the bin size so small that the histogram would approximate to a continuous curve (see Fig. 1.4(e)); this could alternatively be viewed as a very good theoretical prediction about the numbers of men of different heights. Again $n(h)$ is to be interpreted in the same way, but now the total number of men appearing in the histogram is $\int n(h)dh$.

Another example of a distribution for a continuous variable has already been shown in Fig. 1.3, where the variable was denoted by t rather than h.

1.4 Mean and variance

In order to provide some simple description of a distribution such as shown in Fig. 1.4 or 1.5, we need measures of the value at which the distribution is centred, and how wide the distribution is. The mean μ and the mean square deviation from the mean σ^2 (also known as the variance) are suitable for this. Thus σ is the RMS (root mean square) deviation from the mean, and is also known as the 'standard deviation' of the distribution.

The quantities μ and σ^2 refer to the true values for the distribution, and are the estimates that we would obtain for the completely impractical case of an infinite number of unbiassed observations. For a finite set of N separate observations such as shown in Fig. 1.4(a), estimates of μ and σ^2 are respectively

$$\bar{x} = \sum x_i/N \tag{1.2}$$

and

$$s^2 = \sum (x_i - \mu)^2/N. \tag{1.3}$$

where the $\sum$ signs mean that we must add up the specified variables for all the N members of the distribution. In general, the true mean μ is not known, and so eqn (1.3) cannot in fact be used to estimate the variance. Instead it is replaced by

$$s^2 = \frac{1}{N-1} \sum (x_i - \bar{x})^2, \tag{1.3$'$}$$

where the factor $1/(N-1)$ is required in order to make s^2 an unbiassed estimator of the variance (as can be proved fairly readily). This means that, if we repeat our procedure of taking samples lots and lots of times (say L times in all), then we expect that the average of s^2 from all these samples gets closer and closer to the correct value σ^2 as L increases.

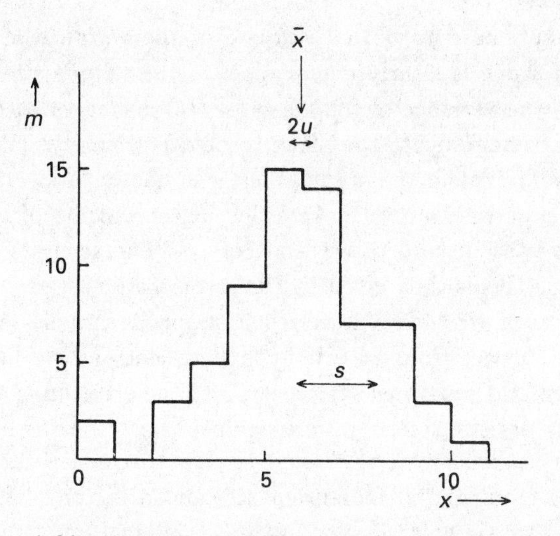

Fig. 1.5. A histogram of a distribution in x. m is the number of entries in each x bin; there are 66 entries in all. The mean $\bar{x}$ and the variance s^2 are estimated as 5.9 and $(2.05)^2$ respectively. The accuracy u to which the mean $\bar{x}$ is determined is smaller than s by a factor of $\sqrt{66}$.

This would not be true if we used a factor of $1/N$ instead of $1/(N-1)$. Another consequence of the $1/(N-1)$ is that one measurement of a quantity does not allow us to estimate the spread in values, if the 'true' value is not known.

Even if we accept the need for the $1/(N-1)$, the definition (1.3′) for s^2 still looks unnecessarily complicated. Since the deviation of an individual measurement from the estimated mean is $(x_i - \bar{x})$, we might have thought that $y = \sum(x_i - \bar{x})/(N-1)$ would have given a simpler estimate of the width of the distribution. The trouble with this is that the definition of $\bar{x}$ is such that $\sum(x_i - \bar{x})$ is guaranteed to be zero, and so y is useless as an estimate of anything. We could instead have used $\sum |x_i - \bar{x}|/(N-1)$, but the modulus sign is rather messy from a mathematical viewpoint, so the definition (1.3′) is adopted.

It is most important to realise that s is the measure of how spread out the distribution is, and is not the accuracy to which the mean $\bar{x}$ is determined. This is known to an accuracy better by a factor of $\sqrt{N}$. Thus by taking more and more observations of x, the variance s^2 will not change (apart from fluctuations) since the numerator and denominator of eqn (1.3) or (1.3′) grow more or less proportionally; this is sensible since s^2 is supposed to be an estimate of the variance of the overall population, which is clearly independent of the sample size N. On the other hand, the variance of the mean (s^2/N) decreases with increasing N; more data help locate the mean to better accuracy. (We return to this point, and explain the origin of the $1/N$ factor, in Section 1.7.1).

In some experiments, we make a few measurements of the quantity we are interested in, and take their average. The accuracy with which we determine this is then given by the error on the mean.

It seems as if we have discovered a simple technique for obtaining an accurate answer from an inferior experiment: all we have to do is to take more and more measurements, and the error on the mean goes down. There are two considerations against this. First the improvement in accuracy is given by $1/\sqrt{N}$, and is slow; to reduce the error by a factor of 10 requires 100 measurements, and a factor of 1000 needs a million repetitions, which in most circumstances is completely impractical. Secondly, it is true that, provided nothing significant changes during the course of this tedious procedure, the statistical error on the mean does decrease as specified. However, all experiments are in danger of having systematic errors as well as random ones. In a well-designed experiment, the systematic error is usually smaller than the random one. Now a repeated set of measurements reduces the statistical error but not the systematic one.* Thus as N increases we reach the point where the systematic error dominates the random error on the mean, and then further repetition of the measurements is of little value. Similarly undetected systematic errors can produce a bias, which again will usually not decrease as the number of measurements increases.

Sometimes the measurements are grouped together so that at the value x_j there are m_j observations (equivalent to $n_j \Delta h$ in Fig. 1.4(d)). Then simple extensions of eqns (1.2) and (1.3′) are

$$\bar{x} = \sum m_j x_j / \sum m_j \tag{1.4}$$

and

$$s^2 = \sum m_j (x_j - \bar{x})^2 / (\sum m_j - 1). \tag{1.5}$$

where the summation now runs over the j bins of the grouped histogram. As usual, s^2 is our estimate of the variance of the distribution; the variance u^2 on the mean is $s^2 / \sum m_j$ (see Fig. 1.5).

* Of course, with extra data, it may be possible to look at potential systematic effects, and to discover how to reduce the error from these sources too, but that would require a lot more thought and work.

For continuous distributions (see, for example Fig. 1.4(e)), these become

$$\bar{x} = \int n(x) x dx / N \tag{1.6}$$

and

$$s^2 = \int n(x)(x - \bar{x})^2 dx / N, \tag{1.7}$$

where

$$N = \int n(x) dx,$$

and where the usual $N - 1$ factor in s^2 has been replaced by N which is assumed to be large for this case.

A minor computational point is worth noting. Eqn (1.3′) can be written

$$s^2 = \frac{N}{N-1}(\overline{x^2} - \bar{x}^2), \tag{1.8}$$

where $\overline{x^2}$ is defined in analogy with eqn (1.2) as

$$\overline{x^2} = \sum x_i^2 / N. \tag{1.9}$$

Thus if someone were reading out the data to you (or if you were accepting the data on a computer), it is not necessary for this to be done twice, first for you to calculate $\bar{x}$, and then to obtain s^2 from eqn (1.3′). Instead $\overline{x^2}$ and $\bar{x}$ can be calculated in a single pass over the data, and then s^2 calculated from eqn (1.8) at the end. If your pocket calculator has the ability to compute standard deviations, it is likely that it does it this way.

However, in using eqn (1.8) for cases where s^2 is small compared with $\overline{x^2}$ and $\bar{x}^2$, it is vital to keep enough significant figures in the numerical calculation, or the obtained value of s^2 can be meaningless. The x values of 9500, 9501 and 9502 provide an example of this.

A final but important point is that, if we attempt to determine the width of a distribution from only a few measurements, our estimate s^2 from eqn (1.3′) will not be very accurate. This is because the particular small sample that we take may have the individual measurements accidentally closer together than might have been expected, or they may be unusually far apart; as the sample size increases, such effects become less likely. In fact, for a set of results that are Gaussian distributed, the fractional error on s is $1/\sqrt{2n-2}$. Thus, for example, with 9 measurements our error estimate is known to only 25%, and there is no sense in quoting more than one significant figure. (See Problem 1.10).

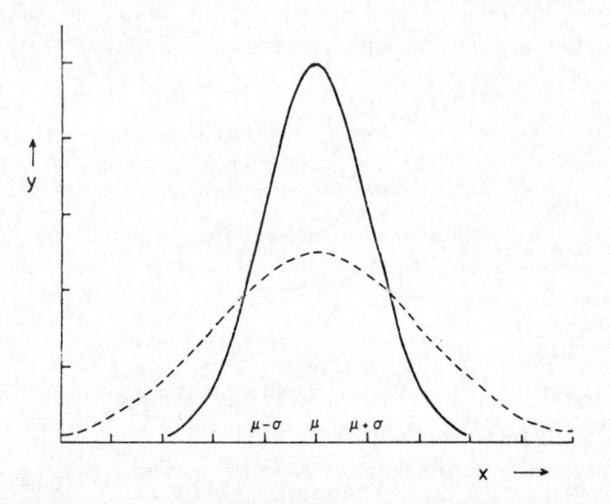

Fig. 1.6. The solid curve is the Gaussian distribution of eqn (1.10). The distribution peaks at the mean μ, and its width is characterised by the parameter σ. The dashed curve is another Gaussian distribution with the same value of μ, but with σ twice as large as the solid curve. Because the normalisation condition (1.11) ensures that the areas under the curves are the same, the height of the dashed curve is only half that of the solid curve at the maxima. The scale on the x axis refers to the solid curve.

1.5 Gaussian distribution

As the Gaussian distribution is of such fundamental importance in the treatment of errors, we now consider some of its properties.

The general form of the Gaussian distribution in one variable x is

$$y = \frac{1}{\sqrt{2\pi}\,\sigma}\exp\{-(x-\mu)^2/2\sigma^2\}. \tag{1.10}$$

The curve of y as a function of x is symmetric about the value of $x = \mu$, at which point y has its maximum value. (See Fig. 1.6.) The parameter σ characterises the width of the distribution, while the factor $(\sqrt{2\pi}\,\sigma)^{-1}$ ensures that the distribution is normalised to have unit area underneath the whole curve, i.e.

$$\int_{-\infty}^{+\infty} y\,dx = 1. \tag{1.11}$$

The parameter μ is the mean of the distribution, while σ has the following properties.

(i) The mean square deviation of the distribution from its mean is σ^2. (This is the reason that the curious factor of 2 appears within the exponent in eqn (1.10). Otherwise the root mean square deviation from the mean would have been $\sigma/\sqrt{2}$, which is unaesthetic.)

(ii) The height of the curve at $x = \mu \pm \sigma$ is $1/\sqrt{e}$ of the maximum value. Since

$$1/\sqrt{e} = 0.607,$$

σ is very roughly the half width at half height of the distribution.

(iii) The fractional area underneath the curve and with

$$\mu - \sigma \leq x \leq \mu + \sigma \tag{1.12}$$

(i.e. within $\pm\sigma$ of the mean μ) is 0.68.

(iv) The height of the distribution at its maximum is $(\sqrt{2\pi}\,\sigma)^{-1}$. As σ decreases the distribution becomes narrower, and hence, to maintain the normalisation condition eqn (1.11), also higher at the peak.

By a suitable change of variable to

$$x' = (x - \mu)/\sigma \tag{1.13}$$

any normal distribution can be transformed into a standardised form

$$y = \frac{1}{\sqrt{2\pi}}\exp(-x'^2/2), \tag{1.14}$$

with mean zero and unit variance.

A feature which helps to make the Gaussian distribution of such widespread relevance is the Central Limit Theorem. One statement of this is as follows. Consider a set of n independent variables x_i, taken at random from a population with mean μ and variance σ^2, and then calculate the mean $\bar{x}$ of these n values. If we repeat this procedure many times, since the individual x_i are random, then $\bar{x}$ itself will have some distribution. The surprising fact is that, for large n, the distribution of $\bar{x}$ tends to a Gaussian (of mean μ and variance σ^2/n). The distribution of the x_i themselves is irrelevant. The only important feature is that the variance σ^2 should be finite. If the x_i are already Gaussian distributed, then $\bar{x}$ is also Gaussian for all values of n from 1 upwards. But even if x_i is, say, uniformly distributed over a finite range, then the sum of a few x_i will already look Gaussian. Thus whatever the initial distribution, a linear combination of a few variables almost always approximates to a Gaussian distribution.

An example of the Central Limit Theorem is given below in Section 1.7.1.

1.6 The meaning of σ

Having concluded our mathematical digressions, we now return to our consideration of the treatment of errors.

For a large variety of situations, the result of repeating an experiment many times produces a spread of answers whose distribution is approximately Gaussian; the approximation is likely to be good especially if the individual errors that contribute to the final answer are small. When this is true, it is meaningless to speak of a 'maximum possible error' of a given experiment since the curve in Fig. 1.6 remains non-zero for all values of x; the 'maximum possible error' would be infinite, and although this would make it easy to calculate the 'error' on any experiment, it would not distinguish a precision experiment from an inaccurate one.

It is thus customary to quote σ as the accuracy of a measurement. Since σ is not the maximum possible error, we should not get too upset if our measurement is more than σ away from the expected value. Indeed, we should expect this to happen with about $\frac{1}{3}$ of our experimental results. Since, however, the fractional areas beyond $\pm 2\sigma$ and beyond $\pm 3\sigma$ are only 5% and 0.3% respectively, we should expect such deviations to occur much less frequently.

The properties of Gaussian distributions are commonly used in interpreting the significance of experimental results. This is illustrated by the following example.

We measure the lifetime of the neutron in an experiment as 950 ± 20 seconds. A certain theory predicts that the lifetime is 910 s. To what extent are these numbers in agreement?

We consult Fig. 1.7, which is a graph showing the fractional area under the Gaussian curve with

$$|f| > r, \tag{1.15}$$

where

$$f = \frac{x - \mu}{\sigma}, \tag{1.16}$$

i.e. it gives (on the right hand vertical scale) the area in the tails of the Gaussian beyond any value r of the parameter f, which is plotted on the horizontal axis. In our example of the neutron lifetime, $f = 2$ and the corresponding probability is 4.6%. Thus if 1000 experiments of the same precision as ours were performed to measure the neutron lifetime, and if our theory is correct, and if the experiments are bias-free, then we expect about 46 of them to differ from the predicted value by at least as much as ours does. Of course, we still have to make up our mind

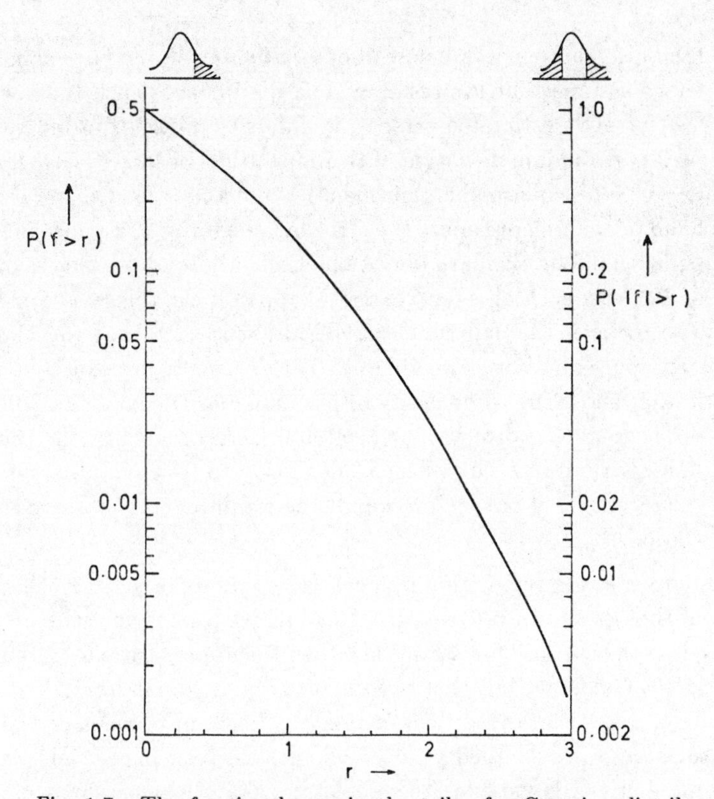

Fig. 1.7. The fractional area in the tails of a Gaussian distribution, i.e. the area with f greater than some specified value r, where f is the distance from the mean, measured in units of the standard deviation. The scale on the left hand vertical axis refers to the one-sided tail, while the right hand one is for $|f|$ larger than r. Thus for $r = 0$, the probabilities are $\frac{1}{2}$ and 1 respectively.

whether we regard the theory (and the experiment) as satisfactory or not, but at least we have a number on which to base our judgement.

In order to rely on the numerical value of this probability, it is essential to ensure that the following assumptions are satisfied.

(i) The value of the quantity of interest has been correctly calculated (e.g. there are no important systematic biasses).

(ii) The magnitude of the error has been correctly calculated. This is particularly important, in that an incorrect estimate of the accuracy of the experiment could have a very large effect on the calculated significance of our result and hence on our conclusions.

(See also the remarks below about *estimates* of σ.) For example, let us assume that the result of an experiment is such that, with the correctly estimated error σ, it differs from the expected value by 2σ. A random deviation of this magnitude or larger is expected in ~5 % of experiments, and hence is not too rare. On the other hand, if we underestimate the error by a factor of 2, we now have a (spurious) four standard deviation effect, whose probability is only $6 \cdot 10^{-5}$ i.e. such an effect 'cannot' happen if the theory is correct.

(iii) The form of the experimental resolution is such that the Gaussian approximation is reasonable. This is usually not exactly true, in that the actual probability of obtaining large deviations (above $\sim 3\sigma$) from the correct value is often underestimated by the Gaussian distribution. This effect is also likely to result in an artificial enhancement of our estimation of the significance of observed deviations.

It is important to realise that σ in eqn (1.16) is supposed to be the true value of the experimental error. If instead all we have is an estimate for σ based on the spread of observations, then the appropriate distribution for f is not Gaussian, but that of Student's t (see Appendix 5).

In most cases, the theoretical estimate y^{th} will have an uncertainty σ' associated with it; theory, after all, is based on experiment, and hence predictions in general are calculated from other measured quantities which of course have their own experimental errors. In that case, we repeat the above procedure of consulting Fig. 1.7, but we redefine f for this case as

$$f = \frac{y^{obs} - y^{th}}{\sqrt{\sigma^2 + \sigma'^2}}, \tag{1.17}$$

where our measured value is $y^{obs} \pm \sigma$. The denominator of (1.17) arises because it is the error on the numerator, assuming that the errors on y^{obs} and y^{th} are uncorrelated (see Section 1.7.1).

Sometimes we are interested in the sign of possible deviations from predicted values.

Example (i)

Is there any evidence for other processes, which we have not allowed for in our theory, contributing to the neutron decay? In other words, is the observed lifetime for neutron decay *smaller* than the predicted value?

Example (ii)

A motorist is accused of speeding by a policeman, who claims that he measured the speed as 38 miles per hour (mph) in an area where the limit was 30 mph. The accuracy of the method used, however, was only ± 5 mph. The relevant question is thus how likely is it that the driver's speed was actually 30 mph *or less*, given the measurement and its accuracy.

In cases where the sign of the possible deviation is of significance, we use the left hand scale of Fig. 1.7, which gives the area of the Gaussian with $f > r$ (or, by symmetry, with $f < -r$), i.e. it is the area in one of the tails.

Fig. 1.7 shows that not only are measurements unlikely to deviate from the correct value by more than a few standard deviations, but also they should not often agree to better than a small fraction of the error. For example, if we measure the neutron lifetime as 909 ± 200 s, we are suspiciously close to the prediction of 910 s. From Fig. 1.7 it can be seen (in principle rather than in practice since the scales are not optimal for very small values of f) that the probability of being within $\frac{1}{200}$ of a standard deviation from a specific value is only 0.4%. Thus

(i) we are unusually lucky on this occasion; or
(ii) our error is over-estimated; or
(iii) we in fact knew the predicted value before we started the experiment, and (perhaps unconsciously) adjusted our measurement to get close to the 'right' answer.

The discussion in this section has been an example of what is known as 'Hypothesis Testing'. We return to a more detailed discussion of this subject in Chapter 2.

1.7 Combining errors

We are frequently confronted with a situation where the result of an experiment is given in terms of two (or more) measurements. Then we want to know what is the error on the final answer in terms of the errors on the individual measurements. We first consider in detail the case where the answer is a linear combination of the measurements. Then we go on to consider products and quotients, and finally we deal with the general case.

1.7.1 Linear situations

As a very simple illustration, consider

$$a = b - c. \tag{1.18}$$

To find the error on a, we first differentiate

$$\delta a = \delta b - \delta c. \tag{1.19}$$

If we were talking about maximum possible errors, then we would simply add the magnitudes of δb and δc to get the maximum possible δa. But we have already decided that it is more sensible to consider the root mean square deviations. Then, provided that the errors on b and c are *uncorrelated*,* the rule is that we add the contributions δb and $-\delta c$ in quadrature:

$$\sigma_a^2 = \sigma_b^2 + \sigma_c^2. \tag{1.20}$$

Two points are worth noting.

(i) If in a particular experiment we know that the measurements of b and c were incorrect by specific amounts δb and δc, then the answer would be incorrect by an amount δa, given in terms of δb and δc by eqn (1.19). But the whole point is that in any given measurement we do not know the exact values of δb and δc (or else we would simply correct for them, and get the answer for a exactly), but only know their mean square values σ^2 over a series of measurements. It is for these statistical errors that eqn (1.20) applies.

(ii) For linear combinations like eqn (1.18), it is the errors themselves which occur in eqn (1.20); percentage errors, which are useful for products (see Section 1.7.2) are here completely irrelevant. Thus if you wish to determine your height by making independent measurements of the distances of your head and your feet from the centre of the earth, each to 1% accuracy, the final answer will not in general be within 1% of the correct answer; in fact, you may well get a result of −40 miles for your height.

Next we discuss why we use quadrature for combining these statistical errors. We look at this in several ways.

(a) Mnemonic non-proof

The errors on b and on $-c$ can be 'in phase' with each other to give

* The meaning of 'uncorrelated' becomes clearer later in this section.

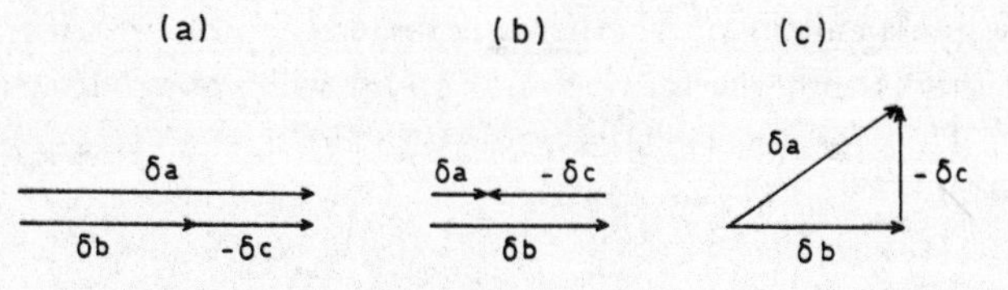

Fig. 1.8. Diagram illustrating the non-proof of formula (1.20). In (a) the contributions from δb and from $-\delta c$ are 'in phase', in (b) they are 'out of phase', while in (c) they appear to be in quadrature.

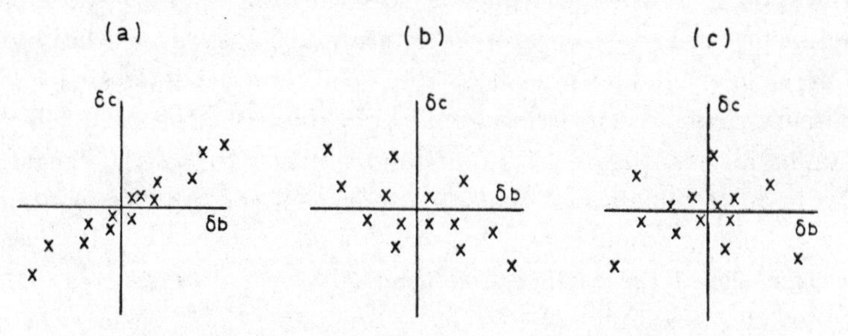

Fig. 1.9. The crosses represent the values of δb and δc for a repeated series of measurements. In (a), these errors are strongly correlated, with $\langle(b-\bar{b})(c-\bar{c})\rangle$ – known as the covariance – being large and positive. The correlation in (b) is less pronounced and slightly negative. In (c) there is almost no correlation, and the covariance is almost zero.

contributions which add up in δa; or they can be 'out of phase', so that they partially cancel in δa. So perhaps on average they are 'orthogonal' to each other and hence Pythagoras' Theorem should be used for obtaining σ_a^2. (See Fig. 1.8.)

We stress that this is not a proof; in particular there is no obvious second dimension in which δb and δc can achieve orthogonality.

(b) Formal proof

$$\begin{aligned}\sigma_a^2 &= \langle[a-\bar{a}]^2\rangle \\ &= \langle[(b-c)-(\bar{b}-\bar{c})]^2\rangle \\ &= \langle(b-\bar{b})^2\rangle + \langle(c-\bar{c})^2\rangle - 2\langle(b-\bar{b})(c-\bar{c})\rangle. \qquad (1.21)\end{aligned}$$

In the above line, the first two terms are σ_b^2 and σ_c^2 respectively. The last term depends on whether the errors on b and c are correlated. In most situations that we shall be considering, such correlations are absent. In that case, whether b is measured as being above or below its average value $\bar{b}$ is independent of whether c is larger than or less than $\bar{c}$. The result of this is that the term $(b - \bar{b})(c - \bar{c})$ will average to zero. (See Fig. 1.9.) Thus for uncorrelated errors of b and c, eqn (1.21) reduces to eqn (1.20).

(c) The infinitesimal probability argument

We perform an experiment which consists of tossing an unbiassed coin 100 times. We score 0 for each heads and 2 for each tails (i.e. the expectation is 1 ± 1 each time we toss the coin). For the complete experiment, we expect on average to score 100. However other scores are possible, including even 0 or 200, so if we were interested in the maximum possible error, this would be ± 100. But the probability of obtaining all heads is only $(^1/_2)^{100}$ (and similarly for tails). Thus if we had a team of helpers such that the experiment could be repeated once every second, we would expect to score 0 or 200 once every $\sim 10^{22}$ years. Since the age of the earth is less than 10^{10} years, we can reasonably discount the possibility of extreme scores, and thus consider instead what are the likely results.

The expected distribution for the final score follows the binomial distribution (see Appendix 3). For 100 tosses, this is very like the Gaussian distribution, with mean 100 and $\sigma \sim 10$. We thus have an example of the Central Limit Theorem mentioned in Section 1.5; by combining a large number N of variables, we end up with something very similar to a Gaussian distribution,* the width of which increases only like $\sqrt{N}$.

(d) Averaging is good for you

We know intuitively that it is better to take the average of several independent measurements of a single quantity than just to make do with a single observation. This follows from the correct formula (1.20), but not from simply adding the errors.

The average $\bar{q}$ of n measurements q_i each of accuracy σ is given by

$$n\bar{q} = \sum_i q_i. \tag{1.22}$$

Then using (1.20) we deduce that the statistical error u on the mean

* Provided, of course, that we don't look at it with too great a resolution, since this distribution is defined only for integral values, whereas the Gaussian is continuous.

is given by

$$n^2u^2 = \sum_i \sigma^2 = n\sigma^2,$$

whence

$$u = \sigma/\sqrt{n}. \tag{1.23}$$

Thus we have obtained the result quoted in Section 1.4 that the error on the mean is known more accurately than the error characterising the distribution by a factor $\sqrt{n}$; this justifies our intuitive feeling that it is useful to average.

The use of the incorrect formula analogous to (1.19) would have led to the ridiculous result that $u = \sigma$. This would imply that the accuracy of the mean is no better than that of a single measurement, and that it would be a waste of effort to repeat an experiment several times in order to determine the mean of the results.

1.7.2 Products

The next simple example is where the answer f is given in terms of the measurements x and y by

$$f = x^a y^b \tag{1.24}$$

where the powers a and b can be positive, negative or fractional. Thus this formula includes simple products, ratios, cubes, square roots, etc. etc.

As in the linear situation, we are going to differentiate this formula, but it is slightly simpler if we first take its logarithm. Then

$$\frac{\delta f}{f} = a\frac{\delta x}{x} + b\frac{\delta y}{y}. \tag{1.25}$$

Again in analogy with our earlier example, after we square and average, provided x and y are uncorrelated, we obtain

$$\left(\frac{\sigma_f}{f}\right)^2 = a^2\left(\frac{\sigma_x}{x}\right)^2 + b^2\left(\frac{\sigma_y}{y}\right)^2. \tag{1.26}$$

That is, the *fractional* error on f is simply related to the *fractional* errors on x and y. This contrasts with the linear case, where *absolute* errors were relevant.

The functions

$$f = xy$$
$$\text{and} \quad f = x/y$$

are so common that it is worth writing the answer explicitly for them as

$$\left(\frac{\sigma_f}{f}\right)^2 = \left(\frac{\sigma_x}{x}\right)^2 + \left(\frac{\sigma_y}{y}\right)^2. \tag{1.27}$$

Thus a 3% error in x and a 4% error in y, assumed as usual to be uncorrelated, would combine to give a 5% error in f.

Because eqn (1.24) is in general not linear in x and y, eqn (1.26) will be accurate only if the fractional errors are small.

1.7.3 The General Case

There are two approaches that can be applied for a general formula

$$f = f(x_1, x_2, \ldots, x_n) \tag{1.28}$$

which defines our answer f in terms of measured quantities x_i each with its own error σ_i. Again we assume the errors on the x_i are uncorrelated.

In the first, we differentiate and collect the terms in each independent variable x_i. This gives us*

$$\delta f = \frac{\partial f}{\partial x_1}\delta x_1 + \frac{\partial f}{\partial x_2}\delta x_2 + \cdots + \frac{\partial f}{\partial x_n}\delta x_n. \tag{1.29}$$

As in our earlier examples, we then square and average over a whole series of measurements, at which point all the cross terms like $\overline{\delta x_1 \delta x_2}$ vanish because the different x_i are uncorrelated. We finally obtain

$$\sigma_f^2 = \sum_{i=1}^{n} \left(\frac{\partial f}{\partial x_i}\right)^2 \sigma_i^2. \tag{1.30}$$

This gives us the answer σ_f in terms of the known measurement errors σ_i. As with products and quotients, if f is non-linear in the x_i, this formula requires the errors σ_i to be small. (See problem 1.6.)

The alternative approach is applicable for any size errors. It consists of the following steps.

(i) Calculate f_o as the value of f when all of the x_i are set equal to their measured values.

(ii) Calculate the n values f_i, which are defined by

$$f_i = f(x_1, x_2, \ldots, x_i + \sigma_i, \ldots, x_n),$$

i.e. where the ith variable is increased from its measured value by its error.

(iii) Finally obtain σ_f from

$$\sigma_f^2 = \sum (f_i - f_o)^2, \tag{1.31}$$

* The curly letter ∂ 's in eqn (1.29) (and later in (1.34)) mean that we should differentiate *partially* with respect to the relevant variable. Appendix 2 contains a brief explanation of partial differentiation.

i.e. we combine in quadrature all the individual deviations caused by moving each variable (one at a time) by its error.

To the extent that the errors are small, this approach should give the same answer as the previous one. For larger errors, the numerical method will give more realistic estimates of the errors. Furthermore, by moving each variable in turn both upwards and downwards by its error, we can deduce upper and lower error estimates for f which need not be identical. Thus, if

$$f = \tan x$$

and

$$x = 88 \pm 1^\circ$$

we obtain

$$f = 29^{+29}_{-10}$$

as compared with

$$f = 29 \pm 14$$

from using eqn (1.30).

When the errors are asymmetric, it is a clear indication that the distribution of f is not Gaussian. Then we should be careful about how we calculate the significance of being, say, two or more errors away from some specified value.

1.8 Systematic errors

In Section 1.9, we shall consider the measurement of a resistance by the method discussed earlier in Section 1.2. We assume that the experiment produced the following results:

$$\left.\begin{aligned} R_1 &= (2.0 \pm 0.1\ \text{k}\Omega)\ \pm 1\%, \\ V_1 &= (1.00 \pm 0.02\ \text{volts})\ \pm 10\%, \\ V_2 &= (1.30 \pm 0.02\ \text{volts})\ \pm 10\%, \end{aligned}\right\} \tag{1.32}$$

where in each case the first errors are the random reading ones, and the second are the possible systematic errors in the various meters.

Although random and systematic errors are different in nature, we may want the overall error estimate as a single figure, rather than expressed separately as above. Then we should add them in quadrature, since reading and calibration errors are uncorrelated. This yields

± 0.1 kΩ, and ± 0.10 V and ± 0.13 V respectively.* When given in this way, however, we lose the distinction between the random and systematic components, which is important, as we shall immediately see.

Next let us consider how accurately we know $V_2 - V_1$. The answer depends on whether the same voltmeter was used to measure both V_1 and V_2, or whether separate meters were employed. In the latter case, presumably there is no correlation between the two $\pm 10\%$ errors on the two readings, and so the linear combination

$$\begin{aligned} V_2 - V_1 &= (1.30 \pm 0.13) - (1.00 \pm 0.10) \\ &= 0.30 \pm 0.16 \text{ V}. \end{aligned} \tag{1.33}$$

In contrast, if the same meter is used, then it is clearly incorrect to assume that the two systematic errors are uncorrelated, since if the first measurement suffers from, say, a -7% calibration error, then so does the second. In this situation, a $\pm 10\%$ systematic uncertainty on each measurement will produce a $\pm 10\%$ systematic error on the result, i.e.

$$\begin{aligned} V_2 - V_1 &= [(1.30 \pm 0.02) - (1.00 \pm 0.02)] \pm 10\% \\ &= 0.30 \pm 0.03 \pm 10\% \\ &= 0.30 \pm 0.04 \text{ V}. \end{aligned} \tag{1.33$'$}$$

It is perhaps a little surprising that for $V_2 - V_1$, which is a linear combination, we are considering *fractional* errors. This is because the uncertainty in the calibration is most simply expressed in terms of a scale factor $f = 1.00 \pm 10\%$. Then the true voltage V^t is given in terms of the measured voltage V^m as

$$V^t = V^m f$$

and the voltage difference

$$V_2^t - V_1^t = (V_2^m - V_1^m)\, f.$$

Thus we are in fact concerned with a product, which explains why fractional errors are relevant.

Alternatively, if the meter had some systematic zero error (which was the same for V_2 and V_1), its effect would exactly cancel in $V_2 - V_1$, and hence its magnitude would be completely irrelevant.

Finally we consider the voltage ratio V_2/V_1. In the case where the same meter was used for V_1 and V_2, its possible scale error of $\pm 10\%$ is irrelevant for V_2/V_1 as it cancels out. In fact it would not matter if

* Throughout this discussion, the magnitudes of errors are rounded to one or two significant figures. In fact only very rarely will it be worth quoting errors to more significant figures. (See Section 1.12).

Table 1.2.
For two measured voltages V_1 and V_2, the table shows the effect of
(i) a random reading error σ_1 or σ_2;
(ii) a common scale error;
(iii) a common zero error
on the voltage difference and the voltage ratio.

	Random error	**Systematic error**	
Quantity	**Reading error**	**Scale error** $1 \pm \delta f$	**Zero error** δz
$V_2 - V_1$	$\sigma^2 = \sigma_1^2 + \sigma_2^2$	$\sigma = (V_2 - V_1)\,\delta f$	$\sigma = 0$
$r = V_2/V_1$	$\left(\frac{\sigma}{r}\right)^2 = \left(\frac{\sigma_1}{V_1}\right)^2 + \left(\frac{\sigma_2}{V_2}\right)^2$	$\sigma = 0$	$\sigma = \frac{V_2+\delta z}{V_1+\delta z} - \frac{V_2}{V_1}$

we were mistaken as to whether the scale was calibrated in millivolts or in volts; or we could read off the voltage in terms of the deflection in centimetres of an oscilloscope beam, without even considering the voltage sensitivity. The voltage ratio V_2/V_1 would not, however, be independent of any zero error (see Table 1.2).

1.9 An example including random and systematic errors

We now work through the case of determining the error on a resistance, whose value is calculated from eqn (1.1), with the measured values as given in eqn (1.32). In fact, the formula for R_2 is best rewritten as

$$R_2 = (V_2/V_1 - 1)\,R_1. \tag{1.1'}$$

The necessary steps in calculating the error on R_2 are then as follows.

(i) Since R_2 is the product of R_1 and $(V_2/V_1 - 1)$, the fractional error on R_2 is determined from the fractional errors on these quantities.

(ii) The absolute error on $V_2/V_1 - 1$ is equal to the absolute error on V_2/V_1.

(iii) The fractional error on V_2/V_1 is determined from the fractional errors on V_1 and V_2.

We now work through these steps in the reverse order, assuming that the same meter is used to measure V_1 and V_2.

(iii) First we need the error on V_2/V_1. Since the systematic scale error cancels, we need consider only the random reading errors on V_1 and V_2, and hence

$$\begin{aligned}\frac{V_2}{V_1} &= (1.30 \pm 0.02)/(1.00 \pm 0.02) \\ &= (1.30 \pm 2\%)/(1.00 \pm 2\%) \\ &= 1.30 \pm 3\% \\ &= 1.30 \pm 0.03.\end{aligned}$$

(ii) The value of $(V_2/V_1 - 1)$ is 0.30 ± 0.03. Thus although this is the same absolute error as that of V_2/V_1, its fractional error is larger.

(i) The value of R_2 is 0.60 ± 0.07 kΩ.

Had the two voltmeters been different, the systematic effects on V_1 and V_2 no longer cancel, and indeed dominate the errors on these voltages. Thus

$$\begin{aligned}V_2/V_1 &= (1.30 \pm 10\%)/(1.00 \pm 10\%) \\ &= 1.30 \pm 0.18.\end{aligned}$$

Then

$$V_2/V_1 - 1 = 0.30 \pm 0.18$$

and

$$R_2 = 0.60 \pm 0.36 \text{ k}\Omega.$$

This is of considerably lower accuracy, because we cannot ignore the calibration uncertainties of the voltmeters.

Instead of working through the combinations of linear functions and of products and quotients, we could have used our derivative formula (1.30) to obtain

$$\sigma^2(R_2) = \left(\frac{\partial R_2}{\partial V_1}\right)^2 \sigma^2(V_1) + \left(\frac{\partial R_2}{\partial V_2}\right)^2 \sigma^2(V_2) + \left(\frac{\partial R_2}{\partial R_1}\right)^2 \sigma^2(R_1) \tag{1.34}$$

where the partial derivatives are evaluated from the equation defining R_2 (eqn (1.1)). Again we would have to be careful about whether the calibration errors on V_1 and V_2 cancelled or not.

We might wonder whether we could have obtained the error on R_2 from eqn (1.1), which consists of the three factors $V_2 - V_1$, R_1 and $1/V_1$. The error on $V_2 - V_1$ we obtained in Section 1.8 earlier; those on R_1 and V_1 are known. Then why cannot we combine the three fractional errors to obtain the fractional error on R_2? The reason is that, regardless of the question of whether V_1 and V_2 are measured by the same meter or not, the error on V_1 cannot be uncorrelated with that of $V_2 - V_1$, since the same measurement V_1 occurs in both. Thus it would be incorrect to use any of the formulae for combining errors which assume the separate components have uncorrelated errors. This is a very important general point: if the same measurement occurs more than once in a formula, it is *wrong* to assume that they can be treated as having independent errors. Both the previous methods of this section avoid this problem.

1.10 Combining results of different experiments

When several experiments measure the same physical quantity and give a set of answers a_i with different errors σ_i, then the best estimates of a and its accuracy σ are given by

$$a = \sum(a_i/\sigma_i^2) \Big/ \sum(1/\sigma_i^2) \tag{1.35}$$

and

$$1/\sigma^2 = \sum(1/\sigma_i^2). \tag{1.36}$$

Thus each experiment is to be weighted by a factor $1/\sigma_i^2$. In some sense, $1/\sigma_i^2$ gives a measure of the information content of that particular experiment. The proof of eqns (1.35) and (1.36) is left as an exercise for the reader (see problem 2.1).

We now give some examples of the use of these formulae.

Example (i)
The simplest case is when all the errors σ_i are equal. Then the best combined value a from eqn (1.35) becomes the ordinary average of the individual measurements a_i, and the error σ on a is $\sigma_i/\sqrt{N}$, where N is the number of measurements. This is all very sensible.

Now there is an entirely different way of estimating the error on the average of the set of results. This consists of simply using the spread of the separate determinations to calculate s (their root mean square deviation from the mean – see eqn (1.3′)), and then the required error is $s/\sqrt{N}$. While this approach ignores the accuracies σ_i with which the individual measurements have been made, eqn (1.36) pays no regard to the degree to which these determinations are mutually consistent. Thus $\sigma_i/\sqrt{N}$ can be regarded as the theoretical error that we expect on the basis of the accuracies of each measurement, while $s/\sqrt{N}$ is an experimental measurement of how spread out are the separate values of a_i.

Which method should we use for determining the error on the mean? If the errors σ_i are estimated correctly and the measurements a_i are unbiassed, then $\sigma_i/\sqrt{N}$ and $s/\sqrt{N}$ should agree with each other satisfactorily. The problem with s is that, especially for small values of N, fluctuations can have a serious effect on its value. Thus some people adopt the strategy of choosing the larger of $\sigma_i/\sqrt{N}$ and $s/\sqrt{N}$ for the error on the mean. My preference is to use $\sigma_i/\sqrt{N}$ provided the two values are reasonably in agreement, and to worry about the consistency of the measurements if s is significantly larger than σ_i.

Discussion of how well the two estimates of the error of the mean should agree, and also the extension of the above example to the situation where the individual errors are unequal, is best dealt with by the χ^2 technique (see Chapter 2).

Example (ii)
The eqns (1.35) and (1.36) are in agreement with common sense when the separate experiments have achieved different accuracies σ_i by using the same apparatus but by averaging different numbers n_i of repeated measurements. In this case the σ_i are proportional to $1/\sqrt{n_i}$ (see Example (d) in Section 1.7.1).

Then the eqns become

$$a = \sum n_i a_i \Big/ \sum n_i \tag{1.37}$$

and

$$N = \sum n_i. \tag{1.38}$$

The first of these says that each of the original measurements of equal accuracy is to be weighted equally; the second is self-evident. (See problem 1.5.)

Example (iii)
We are trying to determine the number of married people in a country, using the following estimates:

$$\begin{aligned} \text{Number of married men} &= 10.0 \pm 0.5 \text{ million}, \\ \text{Number of married women} &= 8 \pm 3 \text{ million}. \end{aligned}$$

Then the total is 18 ± 3 million, where we have obtained the error on the sum as described in Section 1.7.1.

If, however, we assume that the numbers of married men and women are equal, each provides an estimate of half the required answer. Then we use eqns (1.35) and (1.36) to determine the married population as 20 ± 1 million.

Thus we see that calculating the sum or the average of experimental quantities are not equivalent to each other, as the latter implicitly assumes that the quantities we are averaging are supposed to be equal. This extra information results in improved accuracy for the answer.

1.11 Worked examples

We now present some simple worked examples that illustrate the earlier sections of this chapter. The confident student can proceed directly to Section 1.12.

1.11.1 Mean and variance

The following are measurements in grams of the mass of an insect, all with equal accuracy:

5.0 5.3 5.9 5.3 5.2 5.7 5.4 5.1 4.8 5.3

What is our best estimate of its mass, and how accurate is this?

We determine the average of the measurements from eqn (1.2). We first calculate $\sum m_i = 53.0$ g, where m_i are the individual measurements. Then

$$\overline{m} = \sum m_i/10 = 5.3 \text{ g}$$

Next we use eqn (1.3′) to estimate s^2, the variance of the distribution of the measurements. The individual deviations δ_i from the estimated mean of 5.3 g are

−0.3 0.0 0.6 0.0 −0.1 0.4 0.1 −0.2 −0.5 0.0

(As a check, we note that the sum of these deviations is zero, as it must be.) Then

$$s^2 = \frac{1}{9} \sum \delta_i^2 = 0.91/9 = 0.10 \text{ g}^2$$

Thus our estimate of the width of the distribution is $\sqrt{s^2}$, which is 0.3 g. Numerically, this seems not unreasonable when we look at the individual deviations from the mean. Finally the error on the mean is better than this by a factor of $\sqrt{10}$, so we quote our answer as

$$5.3 \pm 0.1 \text{ g}.$$

1.11.2 Using a Gaussian distribution

An experiment measures the current gain of a transistor as 201 ± 12. The expected value is 177 ± 9. Assuming that both these are Gaussian distributed, how consistent are the measurements?

The difference between the measured and expected values is 24, and the error on the difference is $\sqrt{12^2 + 9^2} = 15$. Thus we want to know how likely it is that, if the true difference were zero and we performed an experiment with an accuracy of ± 15, the observed difference would deviate from zero by at least 24. This discrepancy is 1.6 standard deviations, and from tables of the Gaussian distribution we find that the fractional area beyond 1.6σ is 10%. Thus in about 1 experiment in 10, we would expect a random fluctuation to give us a deviation of at least this magnitude, assuming the two values really are perfectly consistent. This probability does not seem too small, so we are likely to be happy to accept the two measurements as agreeing with each other. (We really should have decided beforehand on the cut-off level for an acceptable probability.)

1.11.3 Central Limit Theorem

Illustrate the Central Limit Theorem as follows. Add together four random numbers r_i, each distributed uniformly and independently in the range 0 to 1 (taken from Appendix 7 at the end of the book, or from your pocket calculator), to obtain a new variable $z = \sum r_i$. Repeat the procedure 50 times in all, to obtain a set of 50 z_j values. Plot these as a histogram, and compare it with the appropriate Gaussian distribution.

Using the first 200 random numbers of Appendix 7, we obtain the following z_j values:

1.759 2.161 2.150 2.896........................2.792 0.834

These are drawn as a histogram in Fig. 1.10. Also shown there is the curve

$$y = \frac{50}{\sqrt{2\pi}\,\sigma} \exp\left\{-(z-\mu)^2/2\sigma^2\right\} \Delta z \tag{1.39}$$

where y is the number of entries per bin of width Δz, and the mean μ and width σ are chosen according to the paragraph below eqn (1.14) as 2 and $\sqrt{4/12}$ ($\sqrt{1/12}$ being the RMS of the uniform x distribution of width 1 – see problem 1.2(c) – and 4 being the number of x values added to construct z). Agreement between the histogram and the curve would be expected only if the number of x values added together were

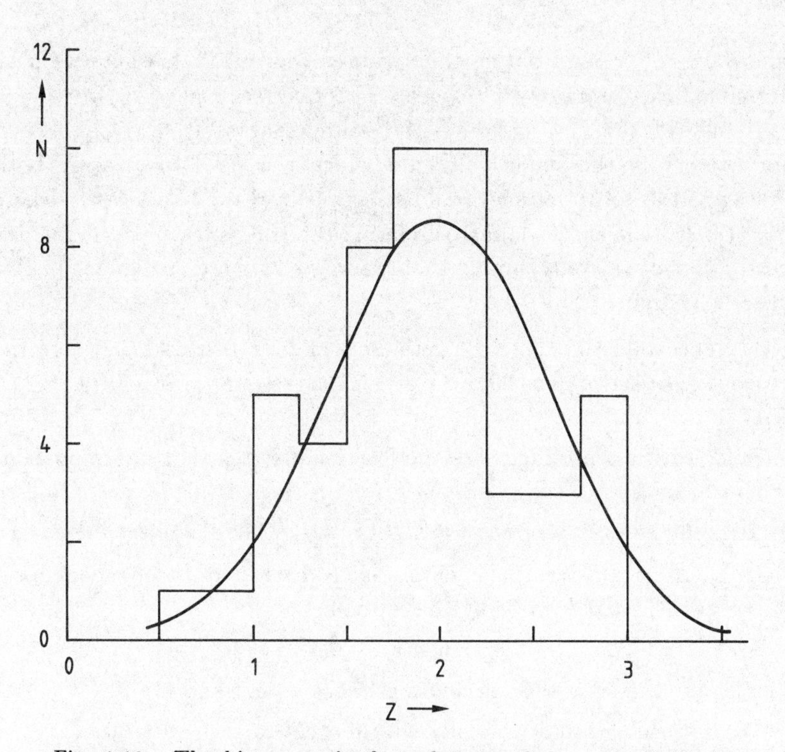

Fig. 1.10. The histogram is that of 50 z values, each of which was obtained by adding four consecutive random numbers from the Table A7.1. The curve is the Gaussian, given by eqn (1.39) of the text. According to the Central Limit Theorem, the distribution of the sum of n random numbers should approximate to a Gaussian if n is large. Even with $n = 4$, and with only 50 entries, the histogram is reasonably consistent with the curve.

large and if there were many entries in the histogram. We see that even with only four x's added, and with their distribution being uniform, the resulting z distribution is reasonably consistent with a Gaussian, and of the expected parameters.

1.11.4 Combining errors

(i) Find the total voltage across two resistors in series, when the voltages across the individual resistors are 10.0 V and 5.0 V, each measured with 10% random error.

Since the total voltage $V = V_1 + V_2$, we need to combine absolute errors rather than fractional ones. We thus express the voltages as 10.0 ± 1.0

and 5.0 ± 0.5 V, and obtain their sum as 15.0 ± 1.2 V, where we have combined the errors by Pythagoras' Theorem (see eqn (1.20)).

If instead the 10% error on the voltages arose from a calibration uncertainty in the single voltmeter which was used to measure both voltages, the total voltage would be given by $V = (V_1 + V_2)f$, where $f = 1.0 \pm 0.1$ is the scale factor relating the true voltage to the meter's reading, and we are assuming that reading errors are negligible. In this case, $V = 15.0 \pm 1.5\ V$.

(ii) The voltage V across a resistor is determined as IR, where the current I is measured as 20 ± 2 mA, and the resistance R is 1.0 ± 0.1 kΩ. Find V.

Since this is a product, it is the fractional errors that are important. These are each 10%, and so the error on the answer is, according to Pythagoras' Theorem, 14% (see eqn (1.27)). Thus V is 20 ± 3 V.

(iii) A certain result z is determined in terms of independent measured quantities a, b and c by the formula

$$z = a \ln c - bc.$$

Determine the error on z in terms of those on a, b and c.

We first differentiate partially with respect to each variable:

$$\frac{\partial z}{\partial a} = \ln c, \qquad \frac{\partial z}{\partial b} = -c, \qquad \frac{\partial z}{\partial c} = \frac{a}{c} - b.$$

Then we use eqn (1.30) to obtain

$$\sigma_z^2 = (\ln c)^2\sigma_a^2 + c^2\sigma_b^2 + (\frac{a}{c} - b)^2\sigma_c^2.$$

(The only part that could give some difficulty is the contribution from the error in c. It is important to combine the two contributions $\frac{a}{c}\delta c$ and $-b\delta c$ before squaring; thus a term $\left\{(a/c)^2 + b^2\right\}\sigma_c^2$ appearing in the answer would be incorrect. In fact, provided a/c and b have the same sign, we get some cancellation between the two contributions, and the error is somewhat smaller than we might at first have thought. If we are still not convinced, we can substitute specific numerical values for a, b and c to obtain z, and then recalculate z for a slightly different value of c, in order to see explicitly the contribution to the change in z arising from a change in c. Provided such changes are small, we should find that $\delta z \sim (\frac{a}{c} - b)\delta c$.)

1.11.5 Combining results

Two independent measurements of the same quantity are 100 ± 5 and 106 ± 7. What is the best estimate?

From the first paragraph of Section 1.10, we find that the relative weights of the two determinations are 49:25~2:1. Thus the weighted average is 102, and its error is given by eqn (1.36) as

$$\begin{aligned}\sigma &= \left(\frac{1}{25} + \frac{1}{49}\right)^{-\frac{1}{2}} \\ &\sim 4.\end{aligned}$$

Thus the best estimate is 102 ± 4. Not surprisingly, the error on the combined answer is smaller than that on either of the individual measurements.

(Had the errors on the two measurements been more different, we would have obtained a significant difference in the errors on the best weighted average and on the simple average. Thus measurements of 100 ± 5 and 94 ± 20 combine to give a best estimate of 100 ± 5, while the simple average is 97 ± 10.)

1.12 Does it feel right?

When you have finally calculated an error, you should spend a short while thinking whether it is sensible.* An important check is that your expression for the error should have the same dimensions as the quantity itself. If not, something terrible has gone wrong with your formula for the error.

Next you should see whether the magnitude of your error agrees with your intuitive feeling about the reliability of your result. Thus if your measurement of the resistance of a coil yields $5 \pm 4\,\Omega$, your error is comparable in magnitude to the quantity itself. This should reflect the fact that your measurement is not significantly different from a value of zero for the resistance. If your feeling is that this experiment really did determine the resistance with reasonable accuracy, then you should go back and look at your calculation of the error again. Incidentally, if the error estimate is correct, it does not mean that your measurement is completely useless. For example, perhaps some behaviour of a circuit incorporating this coil could be understood if its resistance were $25\,\Omega$; your measurement is inconsistent with this.

At the other extreme, an error that is very small compared with the measurement (for example, 1 part in 10^4) suggests a very accurate ex-

* Of course, similar considerations apply to the quantity itself.

periment, and you should check that your results indeed justify such confidence in your answer.*

In a similar way, there is usually little point in quoting your error to a large number of significant figures. There is no need to go as far as calculating the error on the error, but often the input to an error calculation consists of statements like 'I think I can read the deflection of the oscilloscope spot to about half a millimetre'. Now this does not imply that the error on the deflection is 0.5 mm rather than 0.4 or 0.6 mm. Indeed on another day you might have decided that this accuracy was 1/4 mm or 1 mm. Clearly with this degree of arbitrariness concerning the basic accuracies, the error on the answer is unlikely to justify more than one or at most two significant figures. Alternatively, if the error is estimated from the spread of the individual results, we need a large number of repetitions in order to make our error estimate accurate (see last paragraph of Section 1.4).

It is also very important to remember that statistics can provide you with a set of formulae to use, but in an actual practical situation it is not simply a case of choosing the correct formula and applying it blindly to your data. Rather you have to make specific judgements. For example, you may have made several measurements over a period of time, and want to combine them. Then it is necessary to decide whether all the measurements should be included in the average or whether some of them should be discarded because of possible bias; whether all the results have the same accuracy; what are the possible systematic effects; whether there might be a possible time dependence of the answer; etc. Thus although problems in books on errors may have unique answers, real life situations are more interesting and we have to use our experience and intelligence.

Finally, in order to demonstrate that error calculations do present problems even for experienced scientists, Fig. 1.11 shows the way in which the combined best value of the mass of the proton (as obtained from all the different experiments that have measured it) has behaved as a function of time. Assuming that the proton's mass really has not varied, we would expect all these values to be consistent with each other. The fact that they clearly are not demonstrates that either some or all

* A scientist who quoted his error as 1 part in a thousand was asked what the three significant figures represented. He replied 'Faith, Hope and Charity.'

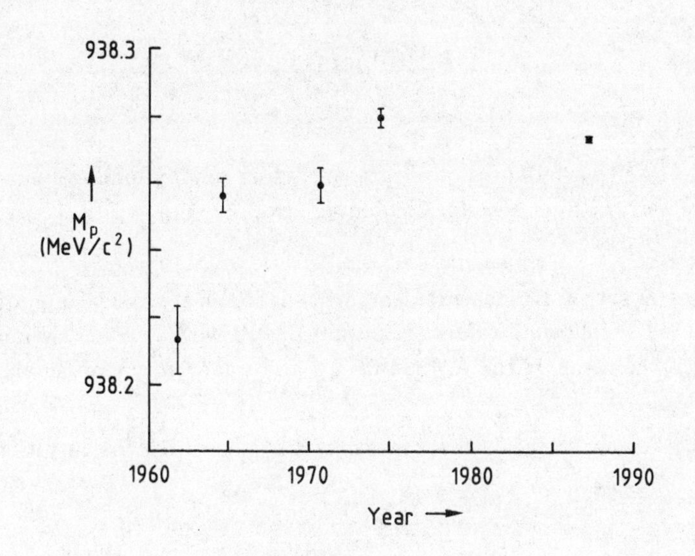

Fig. 1.11. The world average value of the proton mass M_p, as a function of time. The mass is quoted in MeV/c^2. In these units, the electron mass is 0.5109991 MeV/c^2, with an error of 2 in the last decimal place. (Based on information from the Particle Data Group.)

of the errors have been underestimated. Thus, maybe there were biasses present in the experiments which were not allowed for in the quoted errors, or else the statistical errors were for some reason wrong.

Problems

1.1 Write brief notes on errors (including random and systematic errors, and rules for combining errors).

1.2

(i) A set of 13 measurements are made on a physical quantity. The following values are obtained: 0, 1, 2, 3, ..., 12. Estimate the mean $\bar{x}$, the RMS spread s and the accuracy of the mean u.

(ii) A new set of 36 measurements are made with the result that the values

0, 1, 2, ..., 5, 6, 7, ..., 11, 12

occur 0, 1, 2, ..., 5, 6, 5, ..., 1, 0 times respectively.
Estimate $\bar{x}$, s and u.

(iii) The function $n(x)$ is defined as

$$n = \begin{cases} 1/L & \text{for } 0 \le x \le L, \\ 0 & \text{otherwise.} \end{cases}$$

Find the average value of x, and the spread s for this distribution.

(iv) Repeat the problem of (iii) above, but for the function

$$n = \begin{cases} 4x/L^2 & \text{for } 0 \le x \le L/2, \\ 4(L-x)/L^2 & \text{for } L/2 \le x \le L, \\ 0 & \text{otherwise.} \end{cases}$$

(v) Compare the answers for (i) and (iii), and for (ii) and (iv). (You should find that, for a sensible choice of L, the results in (i) and (iii) are approximately the same, and similarly for (ii) and (iv). You should also find that the value of s is smaller for (ii) and (iv), since the measurements are more concentrated near the mean, than are those in (i) and (iii).)

The situation described in (iii) is very relevant for a nuclear physics scintillator, which detects charged particles that pass through it. If the device is of width L in the x direction, all we know when it gives a signal is that a particle passed through it somewhere between $x = 0$ and $x = L$. If we want to specify the x coordinate of the particle (for example, for linking together with other measurements in order to find the direction of the track – see problem 2.4), then we would quote the average and the spread that you have calculated.

1.3 The probability $P(t)\delta t$ of the decay of a radioactive particle between times t and $t + \delta t$ is given by

$$P(t)\delta t = \frac{1}{\tau} e^{-t/\tau}\, \delta t$$

where τ is a constant known as the mean lifetime. Prove the following.

(i) $P(t)\delta t$ behaves like a probability in that its integral over all positive values of t is unity.

(ii) The expected mean value of the decay time t, according to the above probability distribution, is τ (which is why τ is known as the mean lifetime).

(iii) The expected root mean square deviation of decay times about the mean lifetime (i.e. $\sqrt{\langle (t-\tau)^2 \rangle}$) is τ.

Several observations are made of the radioactive decay of a charmed meson. The measured decay times, in units of 10^{-12} seconds, are 0.28, 0.02, 0.09, 0.17, 0.10, 0.62, 0.48, 0.06, 0.85 and 0.08. Use the result (ii) above to obtain an estimate of the lifetime τ of this particle. Given that you know from (iii) above that each individual decay time has an error τ to be assigned to it, what is the error on the estimate of the lifetime that you have just obtained? As an alternative, use the observed scatter of the individual decay times in order to calculate the error on the mean lifetime.

1.4 By measuring yourself with four different rulers, you obtain the following estimates of your height: 165.6 ± 0.3, 165.1 ± 0.4, 166.4 ± 1.0 and 166.1 ± 0.8 cm. What is the best estimate of your height, and how accurate is it? What would have been the best estimate if you had neglected the accuracies of the individual measurements?

1.5 Three schoolchildren *A*, *B* and *C* perform a pendulum experiment with the same apparatus in order to determine the acceleration due to gravity g. An individual measurement consists of timing 100 swings of the pendulum, and this is what *A* does. However, *B* does this twice and averages the two values to obtain an improved answer, while *C* takes the average of ten sets of swings. If *A*'s answer has an uncertainty σ_a, what are the expected accuracies of B's and of C's determinations? (Assume that the dominant error is the random one associated with timing the swings.)

The teacher now takes the three students' determinations ($a \pm \sigma_a, b \pm \sigma_b$ and $c \pm \sigma_c$) and uses the prescription (1.35) and (1.36) to obtain his estimate of g and its error. Show that these are identical with what the teacher would have obtained by taking all 13 individual measurements and averaging them, without regard to which student had performed which determination.

1.6 We wish to determine the ratio f of the strengths of two radioactive sources. For the first we observe 400±20 decays in a minute, and for the second 4±2 in the same time. According to eqn (1.27), the value of f is 100 ± 50. Is this realistic, or is there a way of quoting f and its error which gives a better idea of our knowledge of the ratio?

1.7 For $f = x - 2y + 3z$ (with x, y and z having uncorrelated errors), prove from first principles that

$$\sigma_f^2 = \sigma_x^2 + 4\sigma_y^2 + 9\sigma_z^2.$$

1.8 In each of the following cases, determine the answer and its error, assuming that the errors on the relevant quantities involved in the calculation are uncorrelated.

(i) Determine the distance between the points (0.0±0.2, 0.0±0.3) and (3.0 ± 0.3, 4.0 ± 0.2), and the angle that the line joining them makes with the x axis.

(ii) The number N of particles surviving a distance x in a medium is given by $N_o \exp(-x/\lambda)$, where N_o is the number at $x = 0$, and λ is the mean free path. What is N if $N_o = (1000 \pm 5) \cdot 10^6$, $x = 1.00 \pm 0.01$ m and $\lambda = 0.25 \pm 0.06$ m?

(iii) A particle travels along a straight line trajectory given by $y = a + bx$. If $a = 3.5 \pm 0.3$ cm and $b = (5.0 \pm 0.1) \cdot 10^{-2}$, what is the value of y at (a) $x = 4$ m and (b) $x = 4.0 \pm 0.1$ m?

(iv) The molar specific heat c of a metal at low temperature T is given by $c = aT + bT^3$. If $a = 1.35 \pm 0.05$ mJ mol^{-1} K^{-2}, $b = 0.021 \pm 0.001$ mJ mol^{-1} K^{-4}, and $T = 5.0 \pm 0.5$K, what is the value of c?

1.9 A man lives in a rectangular room for which he wants to buy carpet and wallpaper. The required quantities of these will be proportional to the floor's area and perimeter respectively. He thus measures the floor, and finds that its dimensions are $l \pm \sigma_l$ and $b \pm \sigma_b$, with the errors being uncorrelated. Find the errors on the area and on the perimeter, and show that they are correlated.

This illustrates a general way in which correlations can arise: we make two or more uncorrelated measurements, and then derive new variables which are functions of the original measurements. Other examples include (i) measuring the x and y coordinates of a point, and then calculating the polar variables r and θ; (ii) measuring x and y, and then rotating the coordinate system to obtain x' and y'; and (iii) deducing the intercept and gradient of a straight line fit to several (x, y) sets of data (see Chapter 2, especially Fig. 2.4).

1.10 A measurement with some apparatus produces an answer x that is equally likely to be anywhere in the range 10 to 11. We would say that the likely result μ was 10.5 with an RMS spread σ of $1/\sqrt{12}$ (see problem 1.2(iii)).

Now imagine taking three measurements with this apparatus. You can simulate this by using three random numbers in the range 0 to 1 (which you can obtain either from your calculator, or from a table of random numbers such as is given in Appendix 7), and adding 10 to each. Then calculate $\bar{x}$ and s^2, the estimates of the mean and the variance, from eqns (1.2) and (1.3′). Repeat this procedure several times, and make a list of the $\bar{x}$ and s^2 values. Note that $\bar{x}$ and s^2 scatter about their true values μ and σ^2 respectively. (Compare comments at the end of Section 1.4.)

2

Least squares fitting

2.1 What are we trying to do?

In this chapter we are going to discuss the problem of obtaining the best description of our data in terms of some theory, which involves parameters whose values are initially unknown. Thus we could have data on the number of road accidents per year over the last decade; or we could have measured the length of a piece of metal at different temperatures. In either of these cases, we may be interested to see (i) whether the data lie on a straight line, and if so (ii) what are its gradient and intercept (see Fig. 2.1).

These two questions correspond to the statistics subjects known as Hypothesis Testing and Parameter Fitting. Logically, hypothesis testing precedes parameter fitting, since if our hypothesis is incorrect, then there is no point in determining the values of the free parameters (i.e. the gradient and intercept) contained within the hypothesis. In fact, we will deal with parameter fitting first, since it is easier to understand. In practice, one often does parameter fitting first anyway; it may be impossible to perform a sensible test of the hypothesis before its free parameters have been set at their optimum values.

Various methods exist for parameter determination. The one we discuss here is known as least squares. In order to fix our ideas, we shall assume that we have been presented with data of the form shown in Fig. 2.1, and that it corresponds to some measurements of the length of our bar y_i^{obs} at various known temperatures x_i. Thus the subscript i

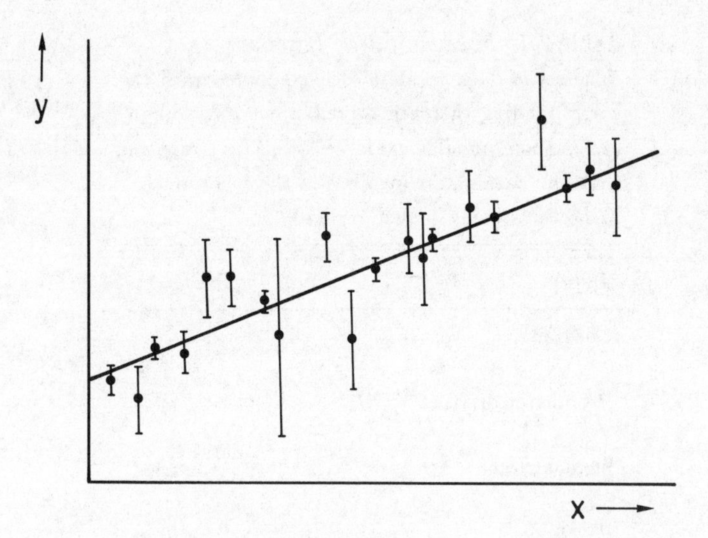

Fig. 2.1 Least squares straight line fit. The data consist of a series of points $(x_i, y_i \pm \sigma_i)$, whose x coordinates are assumed to be known exactly, but whose y coordinates have (varying) experimental uncertainties σ_i. The problem is to find that line such that the sum of the weighted squares of its deviations from all the points is smaller than that for any other possible line. The deviations are measured not as the shortest distance from each point to the straight line, but simply in the y direction. The weighting factor for any point is inversely proportional to the square of its error σ_i; this ensures that the less well measured points do not significantly pollute the better data, while still not being completely ignored in the fit.

labels the different measurements. Each of the length measurements has a certain random error δy_i associated with it; these need not all be the same. On the other hand, the temperatures x_i are assumed to be known exactly.

The theory must be such that, if the parameters we are trying to determine are specified, then there is a unique prediction for y at each of the x_i values. We denote this by $y_i^{th}(\alpha)$, where α is the one or more parameters involved in our theory. Some examples are provided in Table 2.1. Although the method of least squares is general and can be applied to any of these functions, we are going to concentrate on using it to fit straight lines.

Table 2.1. *Possible fitting functions*

The set of data points y^{obs} is compared with the corresponding theoretical predictions y^{th} via eqn (2.1). Some possible examples of $y^{th}(x)$ are given, with the parameters involved in the theoretical predictions being shown explicitly.

Type	y^{th}	Parameters
Constant	c	c
Proportionality	mx	m
Straight line	$a + bx$	a, b
Parabolic	$a + bx + cx^2$	a, b, c
Inverse powers	$a + b/x + \cdots$	$a, b, \ldots$
Harmonic	$A \sin k(x - x_0)$	A, k, x_0
Fourier	$\sum a_n \cos nx$	$a_0, a_1, a_2, \ldots$
Exponential	$Ae^{\lambda x}$	A, λ
Mixed	$\begin{Bmatrix} F_1(x, \alpha_1), x \leq c \\ F_2(x, \alpha_2), x > c \end{Bmatrix}$	α_1, α_2, c

2.2 Weighted sum of squares

If we imagine drawing a whole series of straight lines on the graph of our results (see Fig. 2.2), our judgement of how well any one describes our data would be based on how close it passes to the individual points. The quantity that we use as the numerical quality factor for each line is the weighted sum of squares:

$$S = \sum_i \left(\frac{y_i^{th}(a, b) - y_i^{obs}}{\sigma_i} \right)^2 \tag{2.1}$$

where $y_i^{th}(a, b)$ is the theoretical predicted value at the given x_i, for the

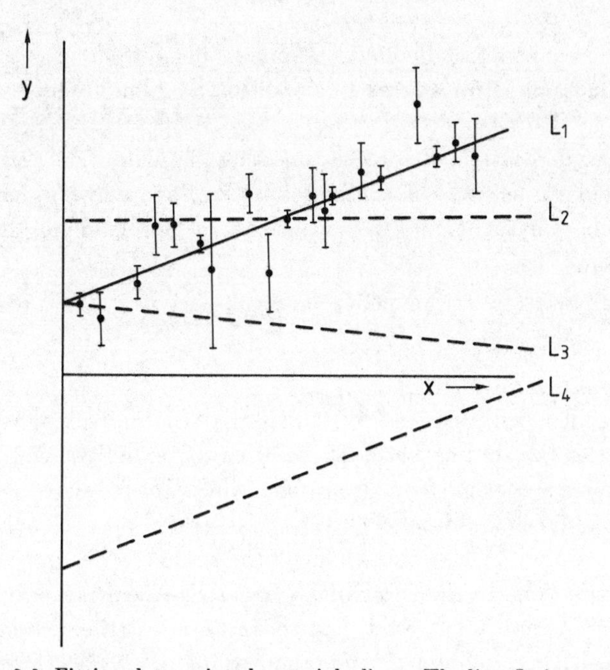

Fig. 2.2 Fitting data points by straight lines. The line L_1 is reasonably close to all the data points (i.e. the deviations are of comparable magnitude to the individual errors), and so gives a small value for S of eqn (2.1). The other lines have large deviations from some or all of the data points, and give large values for S. The best line is defined as that which has the smallest S.

particular values of the parameters a and b for this line, i.e.

$$y_i^{th} = a + bx_i. \tag{2.2}$$

The summation extends over our data points.

The σ_i is some error for each point. In principle it is supposed to be the theoretical error, i.e. the error that would have been expected for the data, assuming it agreed with the theoretical prediction. In practice, we tend to use the observed experimental error on the points (i.e. δy_i), on the grounds that

(i) it makes the algebra of determining the best line very much simpler, and

(ii) provided the points are not far from the line, the two types of errors should not differ greatly.

Clearly the closer each of the y_i^{th} is to the corresponding y_i^{obs}, the smaller S will be. Indeed S is zero if theory and experiment are in perfect agreement for all the data points. Thus we are going to assess the goodness of fit for any particular theoretical line by how small S is (see Section 2.8).

The usual reaction of someone meeting the definition (2.1) for the first time is to ask why it contains a square, and why we need the σ_i^2 in the denominator. Could we not instead use, for example, the simpler expression

$$S' = \sum_i (y_i^{th}(a,b) - y_i^{obs}) \tag{2.3}$$

which is also zero when $y_i^{th} = y_i^{obs}$?

The trouble with expression (2.3) is that the individual terms contributing to the sum can be positive or negative, and we could equally well obtain a zero value for S' from a line which was far above some of the points, and far below others. Furthermore, we can make S' even smaller than zero simply by choosing a line off the bottom of the page, when all the contributions to the sum will be negative. Clearly a very negative value of S' is not at all good, and so we have lost the correspondence between the best fit and the minimum value of S'. To remedy these defects, we need to stop the individual contributions becoming negative. This we could achieve by writing them as $|y_i^{th} - y_i^{obs}|$, but it is mathematically simpler to use $(y_i^{th} - y_i^{obs})^2$.

Thus we are prepared to accept the need for squared terms, but why do we need the σ_i in eqn (2.1), rather than defining

$$S'' = \sum_i (y_i^{th}(a,b) - y_i^{obs})^2? \tag{2.4}$$

Indeed this expression is sometimes used, and as we shall see the best line obtained via eqn (2.1) is equivalent to that from the alternative definition (2.4) for the special case where all the errors σ_i are equal.

The advantage of eqn (2.1) is that it takes better account of data of different accuracies. Effectively each point is weighted by a factor $1/\sigma_i^2$, so that points with smaller errors are much more important. What our best line aims to do is to ensure comparable magnitudes for the fractional distances of the line from each of our points, where the fractional distance is defined as $(y_i^{th} - y_i^{obs})/\sigma_i$, i.e. it is the number of errors by which the prediction differs from the data. Without this factor of σ_i^2 in the denominator of (2.1), the best line would instead give comparable absolute discrepancies $(y_i^{th} - y_i^{obs})$, and thus would treat points with large errors as if they were as accurate as well-measured data.

A second point of some significance is that eqn (2.1) enables us to go on and see whether the agreement between the theory and the data is satisfactory (see Section 2.8); this is not possible if we instead use eqn (2.4).

Having looked at the ideas involved in the least squares method when we are fitting a straight line to the data, we can now turn to an even simpler situation, and see that we obtain very sensible answers. Let us imagine we have n independent measurements $a_i \pm \sigma_i$ of the same physical quantity, and we wish to combine them in order to extract the best estimate a. The least squares method tells us (see eqn (2.1)) to construct

$$S(a) = \sum \left(\frac{a - a_i}{\sigma_i} \right)^2 \tag{2.5}$$

and to determine a by minimising $S(a)$. This yields

$$a = \sum (a_i/\sigma_i^2) / \sum (1/\sigma_i^2)$$

which is exactly the weighted average we already quoted in Section 1.10. For the special case where all the errors σ_i are equal, we obtain the even simpler result

$$a = \sum a_i/n$$

i.e. a is the average of the separate measurements.

There are two ways of looking at what we have just done. Either we can say that the least squares approach provides a derivation of the weighted average answer of Section 1.10; or we can regard these results as to some extent justifying the use of the weighted least squares approach as a technique for determining parameters.

2.3 Determining the parameters

We are now ready to use eqn (2.1) to help us find the best line. We can imagine doing this as follows. We draw lots and lots of lines rather as in Fig. 2.2, for each of them calculate S, and then choose the best line as the one which gives the smallest value of S.

Clearly this is very inefficient, and it is far preferable to find the minimum of S mathematically. The different straight lines of the previous paragraph are each specified by their a and b, and so the minimum of S

is given by*

$$\left.\begin{aligned} \frac{\partial S}{\partial a} &= 0 \\ \text{and} \qquad \frac{\partial S}{\partial b} &= 0 \end{aligned}\right\} \tag{2.6}$$

where from eqns (2.1) and (2.2)

$$S = \sum_i \left(\frac{a + bx_i - y_i}{\sigma_i} \right)^2 \tag{2.7}$$

(from here on, we drop the superscript '*obs*' on the measured y_i).

It is worth reflecting on the fact that this definition of S involves a, b, x_i and y_i. Normally we would think of a and b as constants, and x and y as variables. Here, however, the situation is reversed. We have a fixed set of data points, specified by constants x_i and y_i, while a and b are the parameters of the lines that we are going to vary.

On differentiating S of eqn (2.7) partially with respect to a and then b, we obtain

$$\tfrac{1}{2}\frac{\partial S}{\partial a} = \sum \left(\frac{a + bx_i - y_i}{\sigma_i^2} \right) = 0 \tag{2.8}$$

$$\text{and} \qquad \tfrac{1}{2}\frac{\partial S}{\partial b} = \sum \left(\frac{(a + bx_i - y_i)x_i}{\sigma_i^2} \right) = 0. \tag{2.9}$$

These are two simultaneous equations for our two unknowns, which yield

$$b = \frac{[1][xy] - [x][y]}{[1][x^2] - [x][x]}, \tag{2.10}$$

where the quantities in square brackets are defined by

$$[f] = \sum \frac{f_i}{\sigma_i^2}. \tag{2.11}$$

Note that the weighted means $\langle f \rangle$ of the same quantities are given by

$$\langle f \rangle = [f]/[1]. \tag{2.12}$$

Eqn (2.10) is a nice compact formula for b. Just to make sure that the notation does not obscure what has to be done, we can write it out more fully as

$$b = \frac{\sum(1/\sigma_i^2)\sum(x_i y_i/\sigma_i^2) - \sum(x_i/\sigma_i^2)\sum(y_i/\sigma_i^2)}{\sum(1/\sigma_i^2)\sum(x_i^2/\sigma_i^2) - (\sum x_i/\sigma_i^2)^2}. \tag{2.10$'$}$$

Then a is determined by rewriting eqn (2.8) as

$$\langle y \rangle = a + b\langle x \rangle. \tag{2.13}$$

* See Appendix 2.

In longhand, this is

$$\sum(y_i/\sigma_i^2) = a\sum(1/\sigma_i^2) + b\sum(x_i/\sigma_i^2). \qquad (2.13')$$

Eqn (2.13) shows that the best line (in the least squares fitting sense) passes through the weighted centre of gravity $(\langle x\rangle, \langle y\rangle)$ of all the data points, with the weighting of each point as usual being inversely proportional to σ_i^2.

We note in passing that the average x coordinate $\langle x\rangle$ is given by weighting the individual x_i values by factors that depend on the errors on the y values. This is because the weighting factor applies to the (x_i, y_i) data point as a whole, and not just to the y_i value itself. Thus if all the points but one have large y errors, their weights will be relatively small, and the weighted centre of gravity (in x and in y) will nearly coincide with the single well-measured point. We cannot use the x errors for weighting, since it is an intrinsic feature of this approach that we regard the x_i values as being determined precisely without error.

2.4 The error on the gradient and intercept

Now that we have obtained a and b, we must determine how accurately we know them. Just as in the case of determining the value of a single parameter from a set of measurements – see Section 1.10 – there are two ways in which this can be done. The first makes use of the errors σ_i on the individual measurements in order to determine the errors on a and b, while the second uses the scatter of the measurements about the fitted line (see Fig. 2.3). As in Section 1.10, we regard the former as in general more reliable. The second method will give much larger errors when the minimum value of S is large. This implies that the observed points deviate significantly from the "best" straight line, and hence that the values of the parameters a and b themselves should be regarded with suspicion.

2.4.1 Using the errors σ_i

This is done most easily if we first transform our x coordinates, so that the previous weighted mean $\langle x\rangle$ is now at the origin, i.e.

$$x' = x - \langle x\rangle. \qquad (2.14)$$

Then the equation of the line becomes

$$y = a' + bx' \qquad (2.15)$$

where a' is the height at the position of the weighted mean $\langle x\rangle$. We

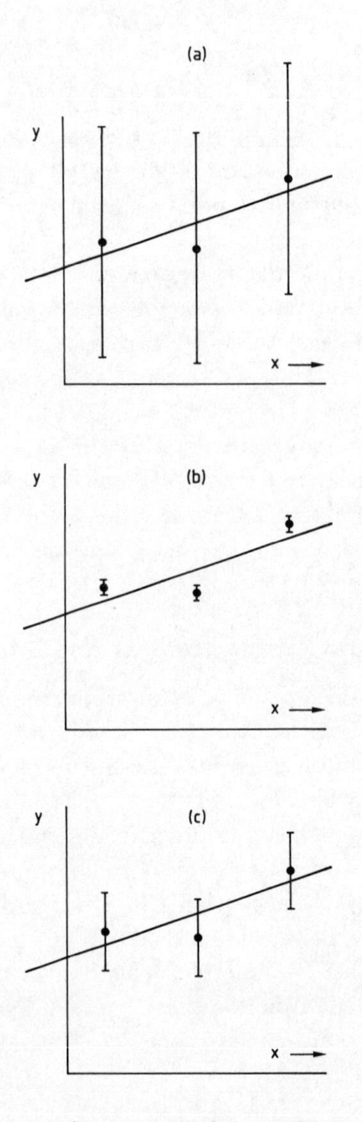

Fig. 2.3 Errors on intercepts and on gradients can be estimated from the assumed accuracies of the individual points (see Section 2.4.1), or from the scatter of the data about the fitted line (Section 2.4.2). For (a) the assumed accuracy method gives larger errors; for (b) this is so for the scatter; in (c), the methods give comparable errors. This is connected with the fact that in (a) the data are (perhaps a bit too) consistent with a straight line, and S_{min} is very small; in (b) S_{min} is large and the fit is improbable; while in (c), $S_{min} \sim n-2$ (see page 60) and the data are reasonably consistent with the line.

do this because the errors on the gradient and on a' are uncorrelated; and we discovered in Chapter 1 that it is much simpler to deal with uncorrelated errors. Incidentally, at $x' = 0$ (i.e. at the weighted average of the x_i values), the estimate for the intercept a' reduces simply to $\langle y \rangle$, the weighted average of the y_i values; this is not surprising.

The values of the errors are given by $(\frac{1}{2}\frac{\partial^2 S}{\partial a'^2})^{-\frac{1}{2}}$ and $(\frac{1}{2}\frac{\partial^2 S}{\partial b^2})^{-\frac{1}{2}}$. Thus we find

$$\left.\begin{aligned} \sigma^2(a') &= 1/[1] \\ \text{and} \qquad \sigma^2(b) &= 1/[x'^2]. \end{aligned}\right\} \tag{2.16}$$

Again we can write these out more fully as

$$\left.\begin{aligned} 1/\sigma^2(a') &= \sum(1/\sigma_i^2) \\ \text{and} \qquad 1/\sigma^2(b) &= \sum(x_i'^2/\sigma_i^2). \end{aligned}\right\} \tag{2.16$'$}$$

Our calculation of $\sigma(a')$ gives us the error on the predicted value at $x' = 0$. If we want the error on the prediction at any other x', then we can propagate the uncorrelated errors on a' and b via eqn (2.15) to obtain

$$\sigma^2(y) = \sigma^2(a') + x'^2\sigma^2(b). \tag{2.17}$$

In particular, if we set $x' = -\langle x \rangle$, the value of y is our original intercept a, and we obtain

$$\sigma^2(a) = \sigma^2(a') + \langle x \rangle^2\sigma^2(b), \tag{2.18}$$

but the errors on a and b are now correlated (unless $\langle x \rangle = 0$). How this correlation arises can be appreciated from Fig. 2.4.

Some simple features of the results we have obtained are worth noting.

(i) The eqns (2.10) and (2.13) for a and b, and (2.16) for the errors, are dimensionally correct.

(ii) If the errors σ_i are all equal, they can be cancelled from eqns (2.10) and (2.13).

(iii) If the errors for all n data points are equal, then

$$\sigma(a') = \sigma/\sqrt{n},$$

i.e. the uncertainty in the intercept at the position of the (weighted) mean is $1/\sqrt{n}$ of that of an individual point.

2.4.2 Using the scatter of points

Here we ignore the errors σ_i, and instead estimate a common error σ for all the measured points by using the scatter of the data about the best fitted line. This method can also be used in the hopefully unusual

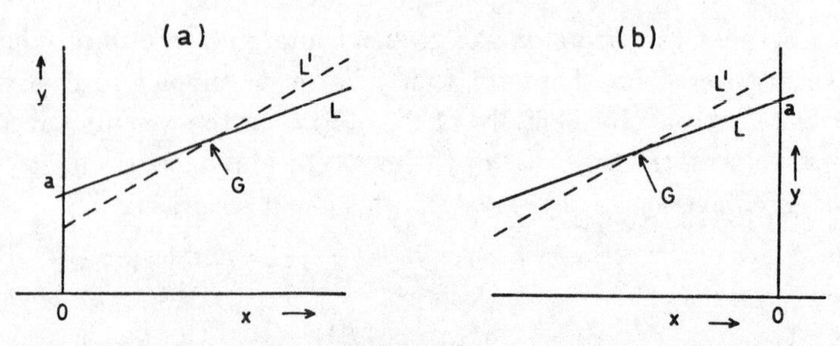

Fig. 2.4 An illustration of the fact that for a least squares straight line best fit, the errors on the gradient b and on the intercept a are correlated, with covariance proportional to $-\langle x \rangle$, the x coordinate of the weighted centre of gravity of the data points (G in the diagrams). The best fit line L passes through G. If the gradient is increased by its error (to give the line L'), then the intercept a will decrease if $\langle x \rangle$ is positive (diagram (a)), or will increase if $\langle x \rangle$ is negative (diagram (b)).

circumstance that the errors on the y values of the individual data points are not known.

We first have to recalculate the values of a and b; since we are changing the values of the errors, the new best line will in general differ somewhat from our previous one. Because all the σ_i are now assumed to be equal, their exact value cancels out in equations (2.10) and (2.13) for a and b. Then we use these new values of a and b in eqn (2.7), to obtain

$$S_{min} = \frac{1}{\sigma^2} \sum (a + bx_i - y_i)^2.$$

The expected value of S_{min} for such a straight line fit to n data points is $n - 2$ (see Section 2.8 below), so we choose

$$\sigma^2 = \frac{1}{n-2} \sum (a + bx_i - y_i)^2. \qquad (2.19)$$

That is, σ is estimated as the root mean square deviation of the measured points (y_i) from the predicted line ($a + bx_i$), except that a factor of $n - 2$ appears in the denominator instead of n. This is in analogy with our estimate of the spread of a series of measurements of a single quantity (see eqn (1.3′)). There the factor was $n - 1$, since we had one unknown parameter – the true value of the quantity being estimated; here we have two, the intercept and the gradient of the straight line.

We now use this value of σ in the right hand sides of eqns (2.16′) to obtain the errors on a' and on b, based on the observed scatter of points about the best line.

2.5 Other examples

So far we have discussed the case of fitting a straight line to the data. Although this is a common situation, our theoretical prediction may consist of some other functional form. What should we do?

One approach is to realise that sometimes a non-linear relation involving two parameters can be transformed in some simple way to be a straight line in different variables. For example, if

$$y = Ae^{\lambda x} \tag{2.20}$$

then $$\ln y = \ln A + \lambda x \tag{2.20'}$$

so that the relationship between $\ln y$ and x is linear. Similarly, if

$$y = ax + bx^3 \tag{2.21}$$

then y/x depends linearly on x^2. Of course in all such cases we must transform our original errors δy_i to those on the new dependent variable (i.e. on $\ln y$ or on y/x in the two examples above).

This technique certainly cannot work if we have more than two parameters in our expression for y, e.g.

$$y = a + bx + cx^2 \tag{2.22}$$

or $$y = \sum a_n \cos nx. \tag{2.23}$$

However, in these cases we simply

(i) substitute the relevant expression for the predicted y as y^{th} in eqn (2.1),
(ii) differentiate partially with respect to each of the p parameters to be determined,
(iii) set these p partial derivations equal to zero, to obtain p simultaneous equations for our p unknown parameters, and finally
(iv) solve these equations to obtain the values of the parameters.

The last stage is straightforward if the function y is such that the parameters occur in it in a linear way, as in eqns (2.22) and (2.23). It is not so, for example, for

$$y = A \sin k(x - x_0)$$

or for $$y = a + x/a.$$

For such non-linear cases, it may well be easiest to find the best values of the parameters simply by varying them until the smallest value of S is found; with a computer this may not be too difficult.

The question of the accuracy with which the parameters are determined for cases more complicated than the straight line fit is most simply dealt with by error matrix techniques, which are beyond the scope of this treatment.

2.6 y^{obs} as numbers

In many applications, the data that we are trying to fit may be such that y is a number of observations. Thus one of the examples in Section 2.1 involved the number of road accidents per year. Alternatively we could plot the number of decaying nuclei of a particular radioactive species as a function of time. Now we will assume that the following conditions are satisfied.

(i) A single event occurs at random and has a fixed probability of occurring during the interval considered.
(ii) The occurrence of one event in a given interval does not influence whether or not another will take place.
(iii) We observe all the events that occur even if they are very close together.

Then under identical conditions a repeated measurement would usually not yield an identical answer, because there are inherent random fluctuations in the processes. The distribution of observed numbers follows what is known as a Poisson distribution,* which is such that the expected root mean square spread of the distribution (when the mean is N) is $\sqrt{N}$. Thus in these circumstances, observed numbers are often quoted as $N \pm \sqrt{N}$.

In these cases, since the error is $\sqrt{N}$, our formula (2.1) reduces to

$$S = \sum_i \left(\frac{(N_i^{th} - N_i^{obs})^2}{N_i^{obs}} \right) \tag{2.24}$$

or

$$S = \sum_i \left(\frac{(N_i^{th} - N_i^{obs})^2}{N_i^{th}} \right) \tag{2.24'}$$

depending on whether we use the observed or the theoretical error for σ_i.

Unfortunately, many people tend to remember eqn (2.24) or (2.24′), and to regard it as the general formula for the weighted sum of squares that can be applied in all cases, even when the observations are not numbers. THIS IS WRONG. Indeed if we replace the N_i in eqn (2.24) by y_i, and try to apply it to the case where y is the length of a metal bar,

* See Appendix 4

we are in effect assuming that the error in the length measurement is $\sqrt{y}$; not only is this dimensionally incorrect, but it is also unrelated to the experimental reality of how accurately each of the lengths is determined.

Thus the best advice is 'FORGET EQN (2.24)'. Now this does sound a bit like the instruction that for the next ten seconds it is absolutely forbidden to think of rhinoceroses. However it causes so many mistakes that it really is important to get into the habit of ALWAYS using eqn (2.1); and if the measurements are numbers that do satisfy the requirements of a Poisson distribution, we simply set $\sigma_i = \sqrt{N_i}$.

2.7 Parameter testing

We now return from the simpler question of what are the best values of the parameters, to the more fundamental one of whether our hypothesis concerning the form of the data is correct or not. In fact we will not be able to give a 'yes or no' answer to this question, but simply to state how confident we are about accepting or rejecting the hypothesis.

There are two different types of hypothesis testing. In the first, our hypothesis may consist simply of a particular value for a parameter. For example, if we believe that the metal we were examining had the special property that it did not expand on heating, we could test that the gradient of the graph of length l against temperature T was consistent with zero. This is parameter testing, which we deal with as described in Section 1.6 earlier.

Thus we assume that all the necessary conditions described there are satisfied and that the error on the parameter has not been obtained from the observed spread of a few measurements. Then we use eqn (1.16) and Fig. 1.7 to tell us how often, if our hypothesis is correct, we would obtain a result that differed from the expected one by at least as much as ours does. If this was suitably low (e.g. less than 5%, or less than 1%), then we would reject our hypothesis. Otherwise, we have no evidence to believe that the hypothesis is wrong (although this is quite a long way from having proved that it is correct).

If the error on our parameter had been estimated from the spread of a few observations, then we should use the relevant Student's t distribution, as explained in Appendix 5.

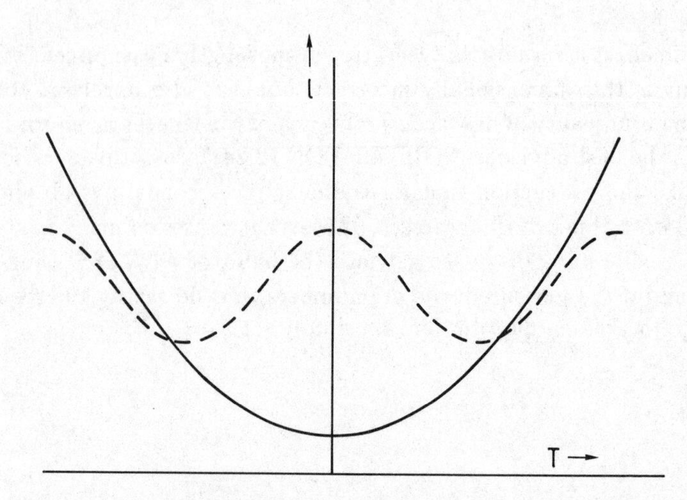

Fig. 2.5 The desirability of examining a distribution rather than simply determining a parameter when we are hypothesis testing. If we fit either the solid or the dashed distribution by eqn (2.25), the resulting value of b is likely to be close to zero. This does not imply that either distribution is constant.

2.8 Distribution testing

In general it is preferable to perform distribution testing. Thus in order to check that a material does not expand on heating, it is more sensible to see whether the graph of l against T is consistent with being constant, rather than simply testing whether a straight line fit gives a gradient close to zero. This is because there are many non-constant distributions which could give a value of $b \sim 0$ if we insisted on fitting an expression

$$l = a + bT \tag{2.25}$$

to the data; a couple of examples are shown in Fig. 2.5.

Distributions are tested by the χ^2 method. When the experimentally observed y_i^{obs} of each experimental point is Gaussian distributed with mean y_i^{th} and with variance σ_i^2, the S defined in eqn (2.1) is distributed as χ^2. So in order to test a hypothesis we

(a) construct S and minimise it with respect to the free parameters,
(b) determine the number of degrees of freedom ν from

$$\nu = n - p \tag{2.26}$$

where n is the number of data points included in the summation for S, and p is the number of free parameters which are allowed to vary in the search for S_{min}, and

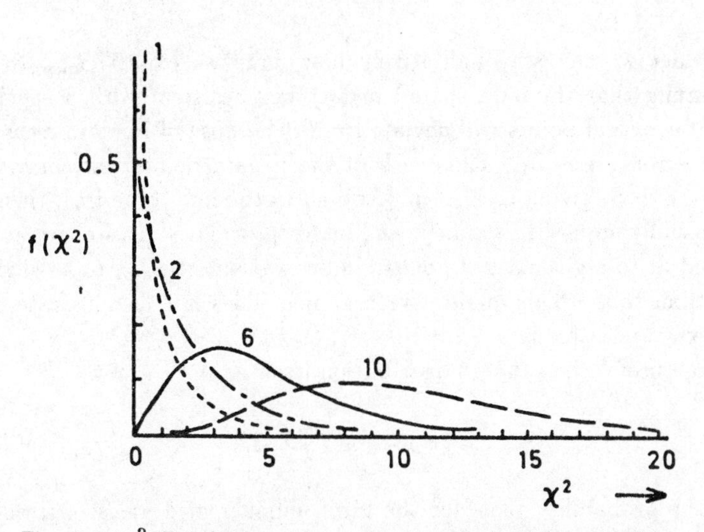

Fig. 2.6 χ^2 distributions, for various numbers of degrees of freedom ν (shown by each curve). As ν increases, so do the mean and variance of the distribution.

(c) look up in the relevant set of tables the probability that, for ν degrees of freedom, χ^2 is greater than or equal to our observed value S_{min}.

In particular we see that, in order to test a hypothesis (for example, that the data are consistent with a straight line), we *must* know the magnitudes of the errors on the individual points. For example, the same set of y_i values appear in Figs. 2.3 (a) (b) and (c); whether they are consistent with a straight line depends on the magnitude of the errors. Without these error estimates, it is impossible to judge their consistency with anything (compare the discussion in Section 1.1).

Some χ^2 distributions, which depend on the number of degrees of freedom, are shown in Fig. 2.6. They have the property that the expectation value

$$\overline{\chi^2} = \nu \tag{2.27}$$

and the variance

$$\sigma^2(\chi^2) = 2\nu. \tag{2.28}$$

Thus large values of S_{min} are unlikely, and so our hypothesis is probably wrong. In this context, 'large' means bigger than $\nu + k\sqrt{2\nu}$, where k is a number like 2 or 3. (Similarly, very small values of S_{min} are also unlikely, and so again something is suspicious – cf. Section 1.6).

In fact we can see qualitatively how large we expect S_{min} to be. Assuming that the data should in fact be consistent with a specified line, the actual points will deviate from it because of random measurement errors by $\sim \sigma_i$. Thus each of the n data points will contribute about 1 to S, giving us $S \sim n$. If we allow the line to be free, then we can usually adjust the gradient and intercept to give us a somewhat improved fit to a given set of points, and so we expect S_{min} to be slightly less than this. (This qualitative argument does not demonstrate that the expected value is $n - 2 = \nu$.)

More useful than the χ^2 distribution itself is

$$F_\nu(c) = P_\nu(\chi^2 > c), \tag{2.29}$$

i.e. the probability that, for the given number of degrees of freedom, the value of χ^2 will exceed a particular specified value c. Some of these are presented in Table A6.3, and shown in Fig. 2.7. The relationship between the χ^2 distribution and that of F is analogous to that between the Gaussian distribution and the fractional area in its tails.

What does F mean? If our experiment is repeated many times, and assuming that our hypothesis is correct, then because of fluctuations we expect a larger value of S_{min} than the particular one we are considering (i.e. a worse fit) in a fraction F of experiments. (The interpretation is thus analogous to that for the case of comparing some standard value with a measured quantity whose error is known, as discussed in the previous section.)

For example, in the situation where we are testing the linearity of the expansion of a metal rod as the temperature is raised, let us assume that there are 12 data points and that when we fit the expression (2.25) to the data, we obtain a value of 20.0 for S_{min}. In this case we have ten degrees of freedom (12 points less the two parameters a and b). From Fig. 2.7, we see that the probability of getting a value of 20.0 or larger is about 3%.

Alternatively, if we were testing the hypothesis that the rod does not expand, then b would be set to zero, the only free parameter is a, and with 12 data points there would be 11 degrees of freedom. In this case S_{min} will be greater than or equal to its value when b was allowed to be a free parameter.

As usual, it is up to us to decide whether or not to reject the hypothesis as false on the basis of this probability estimate, but at least we have a numerical value on which to base our decision.

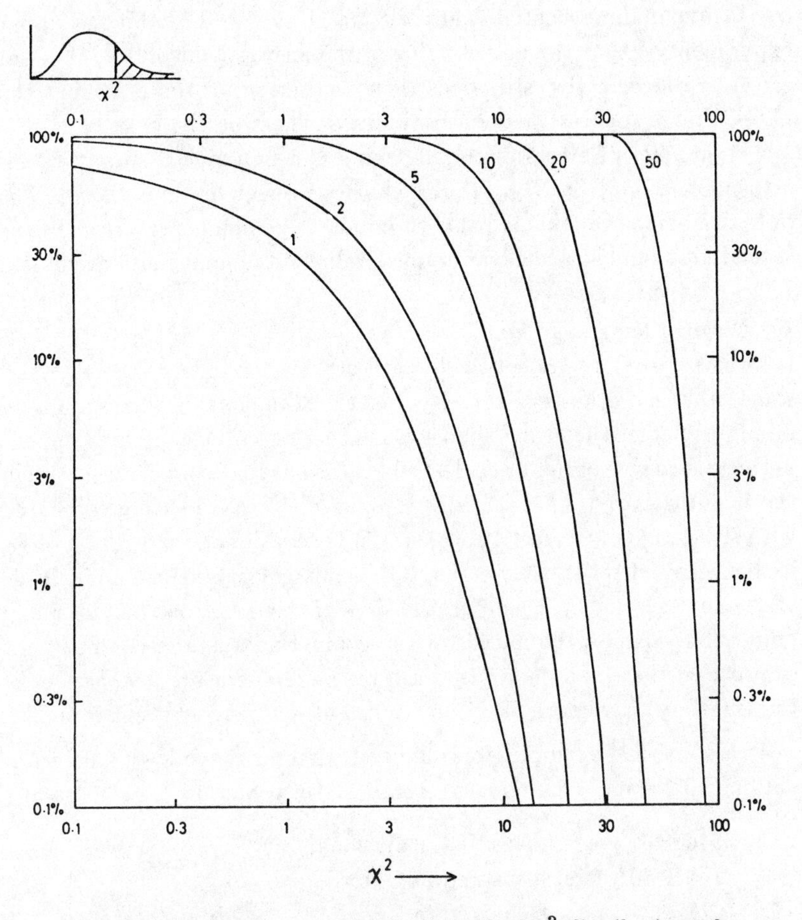

Fig. 2.7 The percentage area in the tail of χ^2 distributions, for various numbers of degrees of freedom, shown by each curve. Both scales are logarithmic. These curves bear the same relation to those of Fig. 2.6 as does Fig. 1.7 to the normal distribution of Fig. 1.6.

In deciding whether or not to reject a hypothesis, we can make two sorts of incorrect decision.

(a) Error of the first kind

In this case we reject the hypothesis H when it is in fact correct. That is, we conclude that our data are inconsistent with eqn (2.25) when in fact the expansion really is linear. This should happen in a well-known fraction F of the tests, where F is determined (from Fig. 2.7) by the maximum accepted value of S_{min}. But if we have biasses in our experiment so that the actual value of the answer is incorrect, or if our errors are incorrectly estimated, then such errors of the first kind can happen more (or less) frequently than we expect on the basis of F.

The number of errors of the first kind can be reduced simply by increasing the limit on S_{min} above which we reject the hypothesis. The only trouble is that this is liable to increase the number of errors of the second kind, and so some compromise value of the limit must be chosen.

(b) Error of the second kind

In this case we fail to reject the hypothesis H when in fact it is false, and some other hypothesis is correct. In our example, this corresponds to failing to detect that the expansion is in fact non-linear. This happens because the value of S_{min} accidentally turns out to be small, even though the hypothesis H (i.e. the theoretical curve y^{th} that is being compared with the data) is incorrect. In general, it is very difficult to estimate how frequent this effect is likely to be; it depends not only on the magnitude of the cut used for S_{min} and on the sizes of the errors on the individual points, but also on the nature of the competing hypothesis. Thus, for example, if there is a quadratic term in the temperature dependence of the length, we are more likely not to detect it if its coefficient is small.

As a numerical example, we could test whether the following determinations of the ages of fragments from the Turin Shroud are consistent:

646 ± 31 years, measured in Arizona,
750 ± 30 years, measured in Oxford,
676 ± 24 years, measured in Zurich.

(These ages, which are quoted in *Nature* **337** (1989), 611, are measured on the radio-carbon age scale, which is not quite the same as the actual age.)

We asssume that the common age is T, and that the individual errors are uncorrelated. We then construct the sum of squares

$$S = \left(\frac{T-646}{31}\right)^2 + \left(\frac{T-750}{30}\right)^2 + \left(\frac{T-676}{24}\right)^2.$$

Our only free parameter is T, and so we minimise S with respect to it. This yields $T = 689$, with the corresponding $S_{min} = 6.4$. Since there are three data points, the number of degrees of freedom $\nu = 3 - 1 = 2$.

From tables of the χ^2 distribution, the probability that $S_{min} \geq 6.4$ for $\nu = 2$ is only about 4%, assuming the relevant hypothesis is true.

That is, provided the fragments are all of the same age, the measurements are unbiassed, and the error estimates are correct, only about one experiment in 25 would give results whose scatter is at least as large as that of the observed results. This is a rather low probability and hence it is not clear that the three measurements are consistent. One possibility is that their experimental errors have been somewhat underestimated.

If we reject the hypothesis because of the large S_{min}, we may well be making an error of the first kind, i.e. we reject the idea of a common age for the fragments, even though this is in fact true. On the other hand, if we accept values of S_{min} as large as this, we are likely to make more errors of the second kind, i.e. we accept the hypothesis in cases where the ages are in fact different. How often this happens depends on factors like the actual difference in ages, and the accuracy of the experiments, as well as the S_{min} cut.

If we are prepared to accept that the results are consistent, we could quote the radio-carbon age from combining the three experiments as 689 ± 16 years, where the error is obtained as $1\Big/\sqrt{\frac{1}{2}\frac{\partial^2 S}{\partial T^2}}$ (see problem 2.1). We see that by combining the answers, we obtain an accuracy that is better than that of any of the individual measurements. However, in view of the somewhat large S_{min}, this procedure may not be justified. In the *Nature* publication, the authors preferred to ignore the errors on the measurements, and to quote the unweighted average as 691 ± 31 years, where the error was calculated from the spread of the measurements.

2.9 Worked example of straight line fit

We here set out a blow-by-blow account of fitting the best straight line to a particular set of data, consisting of the four points as specified by the first two columns of Table 2.2. The third column of the table contains the weight $w_i = 1/\sigma_i^2$ for each data point. We see that the weight for the third point is much smaller than that for the others. Clearly it does not contribute much information to the fit, because of its relatively large error.

The data are plotted in Fig. 2.8(a) (page 67). We now make our eyeball estimate of the best fit, to extract a reasonable value of the gradient and intercept, with which to compare our computed values. In drawing this line, our aim is to minimise $\sum(d_i^2/\sigma_i^2)$, where d_i is the deviation between the data point and the line. Thus the line can miss

Table 2.2. *Worked example of straight line fitting*
The data consists of four sets of $(x_i, y_i \pm \sigma_i)$ values. The weights w_i are $1/\sigma_i^2$. For both types of fit, $x'_i = x_i - \langle x \rangle$, where $\langle x \rangle$ is the relevant average of the x_i; the predicted values y_i^{th} are those for the best fit line; and $d_i = y_i - y_i^{th}$. For the weighted fit, S_{min} is $\sum(d_i^2/\sigma_i^2)$. For the unweighted case, the equation $\sum(d_i^2/\sigma_0^2) = 2$ is used to obtain σ_0.

Data			Weighted fit				Unweighted fit			
x	$y \pm \sigma$	w	x'	y^{th}	d	d^2/σ^2	x'	y^{th}	d	d^2
−2	2.1 ± 0.2	25	−4.325	2.01	+0.09	0.20	−3	1.98	+0.12	0.014
0	2.4 ± 0.2	25	−2.325	2.50	−0.10	0.25	−1	2.41	−0.01	0.000
2	2.5 ± 0.5	4	−0.325	2.99	−0.49	0.96	1	2.84	−0.34	0.116
4	3.5 ± 0.1	100	1.675	3.48	+0.02	0.04	3	3.27	+0.23	0.053
						1.4 = S_{min}				0.183= $2\sigma_0^2$

data points with large errors by a larger amount than for well-measured points. (Compare the remarks at the end of the previous paragraph.)

The first stage of the calculation is the evaluation of the various sums required in eqn (2.10) for the gradient b, and in eqn (2.13) for the intercept a. We obtain

$$\left.\begin{aligned}
[1] &= \sum(1/\sigma_i^2) = 154, \\
[x] &= \sum(x_i/\sigma_i^2) = 358, \\
[y] &= \sum(y_i/\sigma_i^2) = 472.5, \\
[xy] &= \sum(x_i y_i/\sigma_i^2) = 1315, \\
[x^2] &= \sum(x_i^2/\sigma_i^2) = 1716.
\end{aligned}\right\} \tag{2.30}$$

This then gives us the coordinates of the weighted centre of gravity as

$$\left.\begin{aligned}
\langle x \rangle &= [x]/[1] = 2.325, \\
\langle y \rangle &= [y]/[1] = 3.068.
\end{aligned}\right\} \tag{2.31}$$

This is denoted by the star in Fig. 2.8(b).

Next we calculate b from

$$b = \frac{[1][xy] - [x][y]}{[1][x^2] - [x][x]} \tag{2.10}$$

$$= \frac{202510 - 169155}{264264 - 128164}$$
$$= 0.245.$$

The various terms in the numerator and the denominator have been written out in full, so that we can check that there are no serious cancellations between them. If so, we would have to be careful that we were performing our calculations with sufficient precision. Our example is satisfactory from this point of view. Problems could arise, for example, if the x values were spread over a range which was small compared with their average value (e.g. from 6500 to 6501); this could be overcome by redefining the origin of x.

Now we want the intercept. We have

$$\begin{aligned} a &= \langle y \rangle - b\langle x \rangle \qquad (2.13) \\ &= 3.068 - 0.245 \times 2.325 \\ &= 2.50. \end{aligned}$$

Again there appears to be no serious numerical problem. A comparison of a and b with our previous eyeball estimates should be satisfactory.

We can as a check also calculate a directly, rather than via b and eqn (2.13). If we eliminate b from eqns (2.8) and (2.9), we obtain

$$a = \frac{[x^2][y] - [x][xy]}{[x^2][1] - [x][x]}. \qquad (2.32)$$

Substitution of the numerical values of $[x^2]$ etc. then yields $a = 2.50$ again.

Next we want the errors on the fitted quantities. As pointed out in Section 2.4.1, it is useful to calculate first the error on the height $a' = \langle y \rangle$ of the line at the weighted centre of gravity, rather than at $x = 0$. For this we need $[x'^2]$, so we list the values of $x' = x - \langle x \rangle$ in Table 2.2. Then

$$[x'^2] = 884.$$

A useful check is that

$$[x'^2] = [x^2] - [x]^2/[1].$$

We find that our calculated values satisfy this identity.

Then

$$\sigma(b) = 1/\sqrt{[x'^2]} = 0.034$$

and $$\sigma(a') = 1/\sqrt{[1]} = 0.08.$$

If we really want the error on a, we use eqn (2.18):

$$\begin{aligned}\sigma^2(a) &= \sigma^2(a') + (\langle x \rangle \sigma(b))^2 \\ &= 0.08^2 + 0.08^2 \\ &= 0.11^2.\end{aligned}$$

Thus in this case, the error on the intercept at $x = 0$ receives more or less equal contributions from the uncertainty on the overall height of the line, and from that in the gradient.

In Fig. 2.8(b), we show the best fit line, and also the two lines obtained by changing in turn either the intercept or the gradient from their best values by their respective errors. Again we see qualitatively that the error estimates look reasonable.

Finally we want to calculate S_{min} corresponding to our best line. We first evaluate our prediction for each point, i.e.

$$y_i^{th} = 0.245x_i + 2.50, \tag{2.33}$$

and then the deviation

$$d_i = y_i - y_i^{th}.$$

Both y_i^{th} and d_i are displayed in Table 2.2. Then the contributions to S_{min} are d_i^2/σ_i^2, and the final value is

$$S_{min} = 1.4.$$

Since we have four data points and two fitted parameters (a and b), we have $4 - 2 = 2$ degrees of freedom. If our data really do lie on a straight line, S_{min} is expected to follow a χ^2 distribution with two degrees of freedom, whose average value is 2. Thus the observed value of 1.4 is very satisfactory. Assuming that we would rule out values of S_{min} corresponding to 'χ^2 area in the tail' of 5% or lower, any value of S_{min} up to ~ 6 would have been satisfactory (see Table A6.3).

It is worth noting that in predicting the values y_i^{th}, which we require for d_i and then S_{min}, we need the numerical value of the gradient to at least two decimal places. While performing such calculations, we should always ensure that we maintain accuracy at each stage.

We thus present our results as

$$\left.\begin{aligned} b &= 0.24 \pm 0.03, \\ a' &= 3.07 \pm 0.08, \\ a &= 2.50 \pm 0.11, \\ S_{min} &= 1.4 \text{ for two degrees of freedom.} \end{aligned}\right\} \tag{2.34}$$

What would have happened if the errors on the y values had been

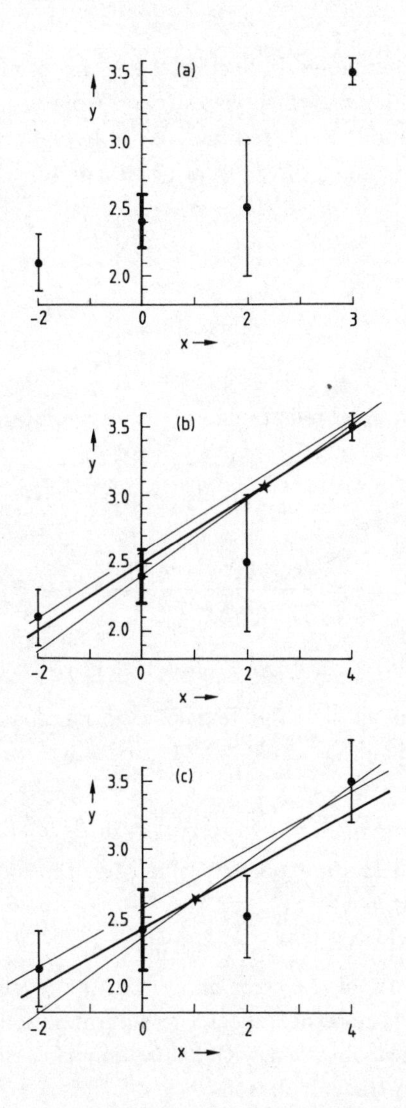

Fig. 2.8 (a) The four data points $(x_i, y_i \pm \sigma_i)$. Readers are invited to cover up (b), and to draw on this diagram their estimate of the best straight line fit to these points. (b) The heavy line is the least squares best fit to the data points, with the individual errors σ_i being taken into account. The best line passes through the weighted centre of gravity of the data points (shown as a star). The two lighter lines are for the intercept or the gradient being increased by one error from their best value. (c) The heavy line is the least squares best fit, when the errors σ_i are ignored. The errors shown on the y values are derived from the scatter of the four points from this best line. The lighter lines are equivalent to those in (b). The star is the location of the unweighted centre of gravity of the data.

unavailable? As explained in Section 2.4.2, we assume that they are all equal to some value σ_0, which we estimate from the observed scatter of the data points about the fitted line. We start, however, by temporarily setting σ_0 equal to unity. Instead of (2.30), we now obtain

$$\left.\begin{aligned} [1] &= 4, \\ [x] &= 4, \\ [y] &= 10.5, \\ [xy] &= 14.8, \\ [x^2] &= 24. \end{aligned}\right\} \quad (2.35)$$

Then the new (unweighted) centre of gravity is given by

$$\left.\begin{aligned} \langle x \rangle &= 1, \\ \langle y \rangle &= 2.625. \end{aligned}\right\} \quad (2.36)$$

The gradient is

$$b = \frac{4 \times 14.8 - 4 \times 10.5}{4 \times 24 - 4 \times 4} = 0.215$$

and the intercept

$$a = 2.625 - 0.215 \times 1 = 2.41.$$

The predictions y_i^{th} and the deviations d_i for this new line are shown in the right hand side of Table 2.2. At this stage, we remember that σ_0 is unknown, and so

$$S_{min} = (\sum d_i^2)/\sigma_0^2 = 0.183/\sigma_0^2.$$

We set this equal to its expected value of 2, the number of degrees of freedom, and obtain

$$\sigma_0 = 0.30.$$

This is our estimate of the error on each of the y values.

This value of σ_0 does not affect the determination of $\langle x \rangle, \langle y \rangle$, a and b that we have just obtained with the assumption that $\sigma_0 = 1$, since it exactly cancels in the calculations.

Finally we are ready to obtain the errors on a and on b. We need $x_i' = x_i - \langle x \rangle$ (see the right hand side of Table 2.2), to calculate

$$[x'^2] = 20.$$

Then $\quad \sigma(b) = 0.30/\sqrt{20} = 0.07$

and $\quad \sigma(a') = 0.30/\sqrt{4} = 0.15.$

Also

$$\begin{aligned} \sigma^2(a) &= \sigma^2(a') + (\langle x \rangle \sigma(b))^2 \\ &= 0.15^2 + 0.07^2 \\ &= 0.16^2. \end{aligned}$$

Thus in this case, the results (see also Fig. 2.8(c)) are

$$\left.\begin{aligned} b &= 0.22 \pm 0.07, \\ a' &= 2.62 \pm 0.15, \\ a &= 2.41 \pm 0.16, \end{aligned}\right\} \tag{2.37}$$

and of course we have no S_{min} with which to make a meaningful test of whether the points lie on the straight line.

By looking at Fig. 2.8(b) and (c), we can compare these results with the earlier ones (eqns (2.34)); we see that the best fit lines are significantly different. For the unweighted fit, all data points are considered equally, even though the third point had a much larger error than the others. Not surprisingly, weighting each point by the correct factor has an effect when the individual errors are not all equal. Clearly we should if at all possible make use of these errors, especially as they also enable us to test the hypothesis that the points do in fact lie on a straight line.

2.10 Summary of straight line fitting

We finally summarise the necessary steps for performing a straight line fit to a series of data points.

(i) Plot the data points on graph paper. This enables us to check that the data looks sensible and to obtain a reasonable estimate of the intercept and gradient.

(ii) Use eqns (2.10) and (2.13) to determine the gradient and intercept.

(iii) Find the weighted mean position $\langle x \rangle$ from eqns (2.12) and (2.11), and then use eqns (2.16) for the errors on the height of the graph at $\langle x \rangle$ and on the gradient.

(iv) Compare the values of the parameters obtained in (ii) and in (i), and also see whether the error estimates look not unreasonable.

(v) Determine the value of S_{min} from eqn (2.7), using the values of a and b from stage (ii).

(vi) Determine the number of degrees of freedom ν from eqn (2.26). Be careful to distinguish between *free* parameters, which are allowed to vary in the fit, and which reduce the number of degrees of freedom; and *fixed* parameters, which do not. Thus a general straight line (eqn (2.2)) has two free parameters, unless either or both of a and b are fixed by the theory, in which case there will be only one or no free parameters respectively.

(vii) Look up the χ^2 table for our number of degrees of freedom in order to find the probability of obtaining a value of S_{min} as big as we observed, or larger.

(viii) If this probability is too small, then realise that the results for a and b are meaningless.

Since the mathematical procedure for fitting straight lines is not only tedious, but also identical in principle for all sets of data, it is well worth writing a computer program to perform it, if you are going to be involved in doing it more than a couple of times. Even with such a tested and bug-free program, however, steps (i) and (iv) must not be omitted.

Problems

2.1 Derive the formula (1.35) for combining two measurements $a_1 \pm \sigma_1$ and $a_2 \pm \sigma_2$ as follows. Assume that they are both consistent with a value $\hat{a}$, and construct an appropriate sum of squares S to test this hypothesis. Then minimise S with respect to $\hat{a}$ to obtain the best value a.

The error on a is given by $\left(\frac{1}{2}\frac{\partial^2 S}{\partial \hat{a}^2}\right)^{-\frac{1}{2}}$. Check that this gives the result (1.36).

Show that you obtain the same result for a by constructing a linear combination $\alpha\, a_1 + (1-\alpha)a_2$ of the measurements a_1 and a_2, with α chosen such that the error on this answer is as small as possible. What is the magnitude of this minimum error?

2.2 Two different measurements of the mass of a given star produce the results of 0.9 ± 0.1 and 1.4 ± 0.2 solar masses. Decide the extent to which they are consistent by the following two methods.

(i) Use a least squares method to test the hypothesis that they are consistent with some arbitrary value, and then look up the relevant probability in χ^2 tables.

(ii) Look up the fractional area in the tails of a normal distribution, for a suitably constructed variable.

2.3

(i) Solve eqns (2.8) and (2.9) for the gradient b, and check that your result agrees with eqn (2.10).

(ii) Use eqn (2.16) for the error on a', to show that it is equal to $\sigma/\sqrt{n}$ for the case where the error on y is σ for all the n measured points.

2.4 In a High Energy Physics experiment a beam of protons is directed from the left onto a small target. The interactions produce secondary particles, whose y coordinates are measured by a series of detector planes accurately situated at x=10, 14, 18, 22, 26 and 30 cm downstream of the target. The first and last planes have an accuracy of $\pm$ 0.10 cm for the y measurements, whilst that for the other four planes is $\pm$ 0.30 cm.

The following y coordinates were obtained for the six planes after one specific interaction: 2.02, 2.26, 3.24, 3.33, 3.92 and 4.03 cm respectively. We want to determine whether these are consistent with lying along a straight track given by $y = a+bx$.

(i) Perform a least squares straight line fit to the data: find the weighted mean x position $\langle x \rangle$, the intercept a, the slope b and the height a' at $x = \langle x \rangle$. Determine S_{min} and the number of degrees of freedom ν, and look up the probability of obtaining a value of S_{min} larger than your value for the given ν.

(ii) Decide whether this track is consistent with coming from the target, which is situated at $x = 0$, and extends from $y = -1$ cm to $y = +1$ cm, as follows. Determine the accuracies with which a' and b are determined, and then use the relation $y = a' + bx'$ (where $x' = x - \langle x \rangle$) to calculate the error ϵ on the extrapolated position y_o of the track at the x' value corresponding to x=0. Then decide whether, given y_o and ϵ, the track is consistent with being produced in the target.

(iii) Calculate the value of S_{min} for the comparison of the straight line $y = 0.1x + 1.0$ cm with the data. (Give yourself a black mark if this is smaller that the S_{min} that you obtained in part (i). Why?) How many degrees of freedom are there?

(iv) How would your answers to part (i) have changed if the accuracies of determining the y coordinate had been taken as ± 0.01 and ± 0.03 cm, rather than ± 0.10 and ± 0.30 respectively, but all other numbers had remained as they were?

2.5 The following data on the number N of observed decays per microsecond (μs) were obtained with a short-lived radioactive source.

$t =$	10	20	30	40	50	60	80	100	(μs)
$N =$	6331	4069	2592	1580	1018	622	235	109	

The expected number of decays is given as a function of time by the relationship $N = N_o \exp(-t/\tau)$, where τ is the lifetime and N_o the number of decays per microsecond at $t = 0$. By fitting a straight line to a graph of $\ln N$ against t, and taking the error for each observed N as being given by $\sqrt{N}$, determine τ and its error.

Appendix 1

Useful formulae

It is very useful to make a short list of formulae summarising each subject you learn. You can then look over this from time to time to remind yourself of the formulae, to ensure you remember what they mean and to what situations they are applicable, etc. If it turns out that you have forgotten, then you can go back to the main text to rectify the situation. Another advantage is that a short list manages to put into perspective the amount of material involved in the topic; if all the formulae can be written on a single page, the total subject matter to learn cannot be too vast. Finally, by making a list yourself, you have to go over the subject to decide what are the important points; this in itself is a valuable exercise.

My own list would be as follows.

(1) For a set of N measurements $\{x_i\}$,

$$\text{Mean } \bar{x} = \sum x_i/N \tag{1.2}$$

$$\text{Variance } s^2 = \sum (x_i - \bar{x})^2/(N-1) \tag{1.3'}$$

$$\text{Error on the mean } u = s/\sqrt{N}$$

(2) Gaussian

$$y = \frac{1}{\sqrt{2\pi}\sigma} \exp\{-(x-\mu)^2/2\sigma^2)\} \tag{1.10}$$

$$\text{Mean } \mu, \qquad \text{variance } \sigma^2$$

(3) Propagating uncorrelated errors

$$a = b \pm c \qquad \sigma_a^2 = \sigma_b^2 + \sigma_c^2 \tag{1.20}$$

$$f = xy \text{ or } x/y \qquad \left(\frac{\sigma_f}{f}\right)^2 = \left(\frac{\sigma_x}{x}\right)^2 + \left(\frac{\sigma_y}{y}\right)^2 \tag{1.27}$$

$$f = f(x_1, x_2, \ldots, x_n) \qquad \sigma_f^2 = \sum \left(\frac{\partial f}{\partial x_i} \sigma_i\right)^2 \tag{1.30}$$

(4) Combining experiments with results $\{a_i \pm \sigma_i\}$

$$a = \sum (a_i/\sigma_i^2) \Big/ \sum (1/\sigma_i^2) \tag{1.35}$$

$$1/\sigma^2 = \sum (1/\sigma_i^2) \tag{1.36}$$

(5) Best fit of straight line $y = a + bx$

$$b = \frac{[1][xy] - [x][y]}{[1][x^2] - [x][x]} \tag{2.10}$$

$$a = \langle y \rangle - b\langle x \rangle \tag{2.13}$$

$$\sigma^2(a') = 1/[1] \tag{2.16}$$

$$\sigma^2(b) = 1/[x'^2] \tag{2.16}$$

where a' is the intercept at $x' = 0$, and $x' = x - \langle x \rangle$

$$S = \sum \left(\frac{a + bx_i - y_i}{\sigma_i} \right)^2 \tag{2.7}$$

$$\nu = n - p \tag{2.26}$$

$$\overline{\chi^2} = \nu \tag{2.27}$$

$$\sigma^2(\chi^2) = 2\nu \tag{2.28}$$

Appendix 2

Partial differentiation

We already know about differentiation in situations where one variable (for example z) is a function of one other (say, x). Thus if

$$z = x \sin x,$$

$$\frac{dz}{dx} = x \cos x + \sin x.$$

If we have functions of more than one variable, ordinary differentiation is replaced by partial differentiation. This involves pretending that, of all the independent variables, only one is allowed to vary at a time, and all others are temporarily kept fixed. Thus if our function is

$$z = x \sin y$$

we can calculate two partial derivatives $\frac{\partial z}{\partial x}$ and $\frac{\partial z}{\partial y}$. (Partial derivatives are always written with curly ∂s, instead of the straight ds of ordinary derivatives). Then for $\frac{\partial z}{\partial x}$, x is treated as a variable, but y is regarded as a constant, so

$$\frac{\partial z}{\partial x} = \sin y.$$

Similarly for $\frac{\partial z}{\partial y}$, x is constant but y is the variable. Hence

$$\frac{\partial z}{\partial y} = x \cos y.$$

In order to help understand what is involved, we can regard z as giving the height of a hilly countryside as a function of the two coordinates x and y, respectively the distances east and north of our starting point. Then $\frac{\partial z}{\partial x}$ is the gradient at any point if we walk along a path going due east over the hillside. Similarly $\frac{\partial z}{\partial y}$ is the gradient as we travel at constant x.

In Chapter 2, we consider the straight line

$$y^{th} = a + bx,$$

where y^{th} is the predicted height of the line at the position x. Now usually we would regard x as the only variable on the right hand side, but since we want to consider lines of all possible gradients and intercepts, in our case a and b are the relevant variables. There are thus three possible partial derivatives $\frac{\partial y}{\partial a}$, $\frac{\partial y}{\partial b}$ and $\frac{\partial y}{\partial x}$.

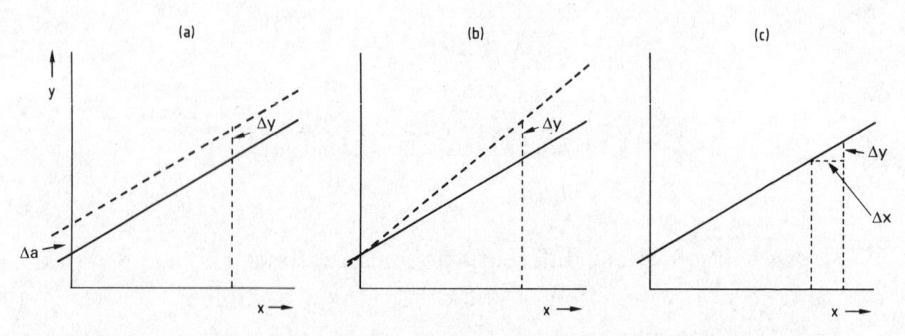

Fig. A2.1 Illustration of the three partial derivatives for the function $y = a + bx$, represented by the solid line. (a) $\frac{\partial y}{\partial a} = 1$ gives the change Δy in y for a change in intercept but for the same gradient, represented by the dashed line. At constant x, $\Delta y = \Delta a$. (b) $\frac{\partial y}{\partial b} = x$ gives the change in y for a change in gradient but with the intercept unchanged. At constant x, $\Delta y = x\Delta b$. (c) $\frac{\partial y}{\partial x} = b$ gives the change in y for a change in x, keeping the line fixed. Then $\Delta y = b\Delta x$.

Thus $\frac{\partial y}{\partial a}$ tells us that rate of change of y with respect to a, at fixed b and fixed x. We obtain

$$\frac{\partial y}{\partial a} = 1.$$

This means that, if we change the intercept of the line by Δa but keep the gradient constant, the change in y at a particular x is $\Delta y = \Delta a$. Similarly

$$\frac{\partial y}{\partial b} = x$$

which means that, if we alter the gradient but not the intercept, the change in y at any x is given by $\Delta y = x\Delta b$.

We do not require it in Chapter 2, but we can also calculate

$$\frac{\partial y}{\partial x} = b,$$

which tells us that for a fixed straight line (i.e. a and b constant), the change in y as we vary x is $b\Delta x$. All three partial derivatives are shown graphically in Fig. A2.1.

Thus the partial differentiation of a function of several variables is no more complicated than the ordinary differentiation of a function of one variable.

Just as the stationary value of a function $z(x)$ requires that the first derivative $\frac{dz}{dx} = 0$, so for a function of two variables $z(x, y)$, it is necessary to have both partial derivatives zero, i.e.

$$\left.\begin{aligned} \frac{\partial z}{\partial x} &= 0 \\ \text{and} \qquad \frac{\partial z}{\partial y} &= 0. \end{aligned}\right\}$$

In general this will not ensure that we have a minimum rather than a maximum or something more complicated. For the case of the χ^2 fit of a line to a set of data points, however, the stationary value that we determine when we set the derivatives of eqn (2.6) equal to zero is the minimum of S.

Appendix 3

The binomial distribution

In Section 1.5 we discussed the Gaussian distribution. In this and the next appendix, we describe the binomial and Poisson distributions.

Let us imagine that we throw an unbiassed die 12 times. Since the probability of obtaining a 6 on a single throw is $\frac{1}{6}$, we would expect on average to end up with two 6's. However, we would not be surprised if in fact we obtained 6 once or three times (or even not at all, or four times). In general, we could calculate how likely we are to end up with any number of 6's, from none to the very improbable 12.

These possibilities are given by the binomial distribution. It applies to any situation where we have a fixed number N of independent trials, in each of which there are only two possible outcomes, success which occurs with probability p, or failure for which the probability is $1 - p$. Thus, in the example of the previous paragraph, the independent trials were the separate throws of the die of which there were $N = 12$, success consisted of throwing a 6 for which the probability $p = \frac{1}{6}$, while failure was obtaining any other number with probability $\frac{5}{6}$.

The requirement that the trials are independent means that the outcome of any given trial is independent of the outcome of any of the others. This is true for a die because what happens on the next throw is completely unrelated to what came up on any previous one. In contrast, if we have four red balls and two black balls in a bag, and consecutively take out two balls at random, the chance of the second being black is influenced by whether the first is red or black; in the former case, the probability is $\frac{2}{5}$, while in the latter it is $\frac{1}{5}$. (On the other hand, if we put the first ball back in the bag before removing the second, the trials would be independent.)

For the general case, the probability $P(r)$ of obtaining exactly r successes out of N attempts is

$$P(r) = \frac{N!}{r!(N-r)!} p^r (1-p)^{N-r}$$

for values of r from 0 to N inclusive. This is because the p^r term is the probability of obtaining successes on r specific attempts, and the $(1-p)^{N-r}$ factor is that for failures on the remaining $N - r$ trials.

But this corresponds to only one ordering of the successes and failures; the combination of factorials then give the number of permutations of r successes and $N - r$ failures. It is important to distinguish between p, the probability of success on a single trial, and $P(r)$, the probability of obtaining r successes out of N trials.

From the equation above, the probability of no 6's in 12 trials is

$$P(0) = \frac{12!}{0!12!}\left(\frac{1}{6}\right)^0\left(\frac{5}{6}\right)^{12} = \left(\frac{5}{6}\right)^{12} = 0.11$$

(remember that 0! is 1). Similarly the probability of one 6 is

$$P(1) = \frac{12!}{1!11!}\left(\frac{1}{6}\right)^1\left(\frac{5}{6}\right)^{11} = 12\left(\frac{1}{6}\right)\left(\frac{5}{6}\right)^{11} = 0.27.$$

The extra factor of 12 arises from the fact that the probability of obtaining a 6 on the first throw, followed by none in the remaining 11 attempts, is $\left(\frac{1}{6}\right)\left(\frac{5}{6}\right)^{11}$. However, in order to obtain just one 6 during the sequence, it could occur on any of the throws, rather than necessarily on the first. Since there are 12 such opportunities, the total probability $P(1)$ is 12 times the probability of obtaining 6 on only the first throw.

In our probability for no 6's, there is no such extra factor, since the only possibility is for every single throw to give a non-6. In contrast, if we ask for two 6's, these could occur in any of the 12 throws, and there are 66 ways this could happen. Our equation correctly gives the probability as

$$P(2) = \frac{12!}{2!10!}\left(\frac{1}{6}\right)^2\left(\frac{5}{6}\right)^{10} = 66\left(\frac{1}{6}\right)^2\left(\frac{5}{6}\right)^{10}$$
$$= 0.30.$$

We can thus use the equation to calculate all the probabilities from $P(0)$ to $P(12)$ for our die example. These are plotted in Fig. A3.1(a). As is necessarily the case, these probabilities add up to unity. This is because, if we throw a die 12 times, we are absolutely sure to find that the number of 6's is somewhere in the range 0 to 12 inclusive.

For the binomial distribution, the expected number of successes $\bar{r}$ is

$$\bar{r} = \sum rP(r)$$

with $P(r)$ as given earlier. A fairly tedious calculation gives

$$\bar{r} = Np$$

which is hardly surprising, since we have N independent attempts, each with probability p of success. Even more boring algebra gives the variance σ^2 of the distribution of successes as

$$\sigma^2 = Np(1-p).$$

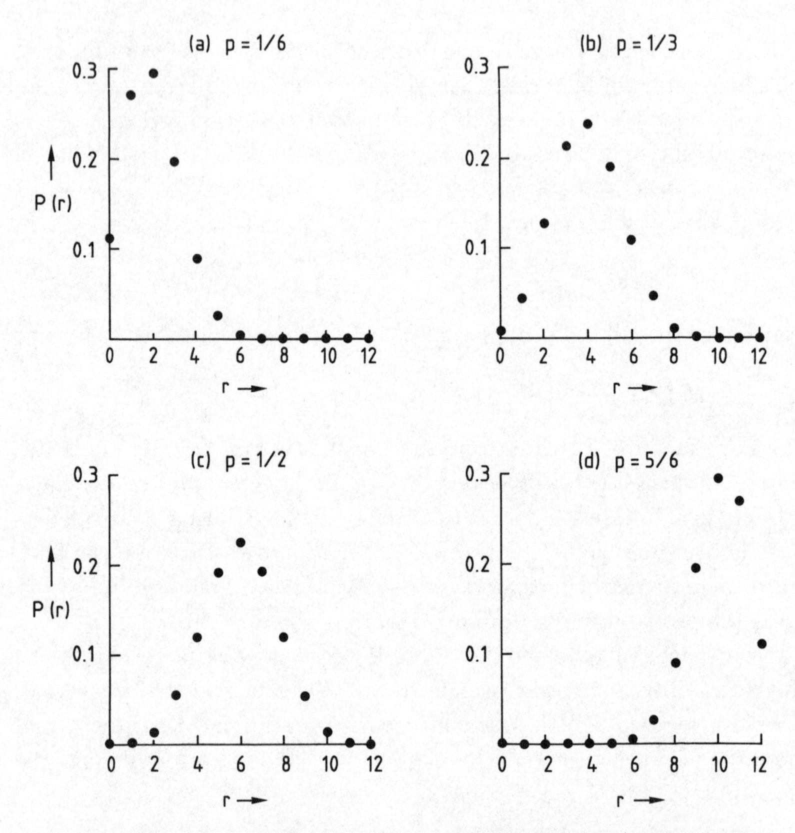

Fig. A3.1 The probabilities $P(r)$, according to the binomial distribution, for r successes out of 12 independent trials, when the probability p of success in an individual trial is as specified in the diagram. As the expected number of successes is $12p$, the peak of the distribution moves to the right as p increases. The RMS width of the distribution is $\sqrt{12p(1-p)}$ and hence is largest for $p = \frac{1}{2}$. Since the chance of success in the $p = \frac{1}{6}$ case is equal to that of failure for $p = \frac{5}{6}$, the diagrams (a) and (d) are mirror images of each other. Similarly the $p = \frac{1}{2}$ situation shown in (c) is symmetric about $r = 6$ successes.

Thus the expected number of successes of our die-throwing experiment was $12 \times (1/6) = 2$, with a variance of $12 \times (1/6) \times (5/6) = 5/3$ (or a standard deviation of $\sqrt{5/3}$). This tells us that we cannot expect that the number of successes will be much larger than a couple of times $\sqrt{5/3}$ above 2, i.e. more than five 6's is unlikely (see Fig. A3.1(a)).

For the same experiment of throwing a die 12 times, we could have changed the rules, and regarded either 5 or 6 as a success. Thus the probability of success is now $2/6 = 1/3$, and our formula for $P(r)$ will give us the new set of probabilities for obtaining any specified number r of successes out of $N = 12$ attempts. These are plotted in Fig. A3.1(b). Also shown in Fig. A3.1 are the values of the probabilities $P(r)$ of r successes out of 12 trials for the cases where $p = 1/2$ (e.g. any even number is a success), and for $p = 5/6$ (e.g. anything but a 1 is a success).

Other examples where we would use the binomial distribution (the second one being approximate) are as follows.

(i) We do 20 experiments in a year. We expect that 95% of the observed results should be within 2 standard deviations of the correct value. How likely are we to have all experiments within 2 standard deviations? Or 19 experiments?

(ii) A large school consists of 55% boys and 45% girls. Two students are chosen at random. What are the probabilities that they are both boys; a boy and a girl; or both girls?

The binomial distribution is thus useful for calculating probabilities of the various possible outcomes of an experiment consisting of a fixed number of repeated trials, provided that the probability of success of an individual trial is known and constant.

Situations in which the number of trials N becomes very large are interesting. If as N grows the individual probability p remains constant, then the binomial distribution becomes like the Gaussian distribution (see Fig. A4.2) with mean Np and variance $Np(1-p)$. On the other hand, if as N grows, p decreases in such a way that Np remains constant, the binomial tends to a Poisson distribution, which we discuss in the following appendix.

Appendix 4

The Poisson distribution

The flux of cosmic rays reaching the Earth is about 1 per cm^2 per minute. So if we have a detector with an active area of $\sim$ 20 cm^2, we might expect on average 3.3 cosmic rays for each 10 second interval. If we actually did the experiment and recorded how many particles were detected in consecutive 10 second intervals, we would clearly never observe 3.3, but would get numbers like 3 or 4, and sometimes 1 or 2 or 5 or 6, and occasionally none or 7 or more. We could produce a histogram showing how many times (n_r) we observed exactly r cosmic rays in our 10 second intervals. By dividing by the total number of intervals T, we could convert n_r into the corresponding probabilities P_r. For large enough T, and assuming certain conditions are satisfied (see below), this distribution of probabilities should approximate to the Poisson distribution with the corresponding average value $\lambda = 3.3$.

The Poisson distribution arises when we observe independent random events that are occurring at a constant rate, such that the expected number of events is λ. The Poisson probability for obtaining r such events in the given interval is

$$P_r = \frac{\lambda^r}{r!}e^{-\lambda} \tag{A4.1}$$

where r is any positive integer or zero. Thus the Poisson distribution has only one parameter λ. Some distributions corresponding to different values of λ are shown in Fig. A4.1.

It is important that the events are independent, which means that the observation of one such event does not affect whether we are likely to observe another event at any subsequent time in the interval.

The cosmic ray situation described above is likely to satisfy these requirements. Cosmic rays seem to arrive at random, and their average rate of arrival is more or less constant. The distribution would not be Poisson if our counter had a significant dead-time, such that it was incapable of recording a cosmic ray that arrived a short time after another, since this would violate the independence requirement. With modern counters and electronics, any dead-time should be very short compared with the average time between cosmic rays, and so again the effect can be ignored.

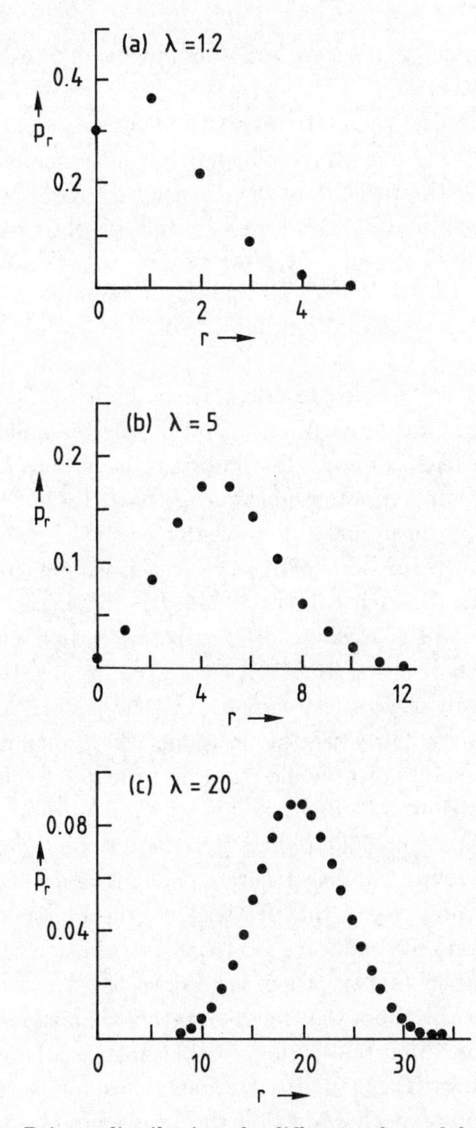

Fig. A4.1 Poisson distributions for different values of the parameter λ. (a) $\lambda = 1.2$; (b) $\lambda = 5.0$; (c) $\lambda = 20.0$. P_r is the probability of observing r events. (Note the different scales on the three figures.) For each value of λ, the mean of the distribution is at λ, and the RMS width is $\sqrt{\lambda}$. As λ increases above about 5, the distributions look more and more like Gaussians.

In a similar way, the Poisson distribution is likely to be applicable to many situations in which we are counting independent events in a fixed time or space interval.

For the probability distribution (A4.1), we can calculate the average value of r, as $\sum rP_r$. This turns out, not too surprisingly, to be λ. We can also calculate the variance of the distribution, as $\sum(r-\lambda)^2 P_r$; this again is λ. Thus, if in a Poisson-type situation we observe N events, we would estimate λ as N, and the error as $\sigma = \sqrt{\lambda} = \sqrt{N}$. This is the basis of the well known $N \pm \sqrt{N}$ recipe that applies to statistical errors in many situations involving the counting of independent events during a fixed interval.

The Poisson distribution also arises as the limit of the binomial distribution. That is, for large N and small p, the binomial distribution looks very much like a Poisson distribution of mean $\lambda = Np$.

Thus if we have a very intense beam of B particles incident on a thin target, such that the probability p of any one of them interacting is very small, then the number r of observed reactions in fact is binomially distributed (B trials, each with probability p of "success"). However, for small p, this will give values of P_r that are indistinguishable from those of a Poisson of mean Bp.

In a similar way, if we select samples of 1000 people, and count the number r with some fairly rare blood group, the probability P_r of obtaining different values of r should again in principle be determined by the binomial distribution. But for any practical purpose, the Poisson distribution with mean $1000p$ (where p is the probability of having the specified blood group) will give accurate enough values of P_r.

A final useful property is that for large λ, the Poisson probabilities are very like those obtained for a Gaussian distribution of mean λ and variance λ. Thus in this situation the values of P_r can be calculated somewhat more easily from the Gaussian rather than the Poisson distribution. Even more important is the fact that many statistical properties are much simpler for Gaussian distributions. Thus, for example, the sum S_{min} of eqn (2.7) is expected to follow the χ^2 distribution provided that the errors σ_i are Gaussian distributed. If we in fact have a situation where the distribution is expected to be Poisson (e.g. for the number of particles scattered through a given angle in a particular nuclear physics experiment), then provided that the number n of such events is large enough, the approximation to a Gaussian will be adequate. In this context, n larger than about 5 is usually taken as sufficient.

The relationship among the binomial, Poisson and Gaussian distributions is shown in Fig. A4.2.

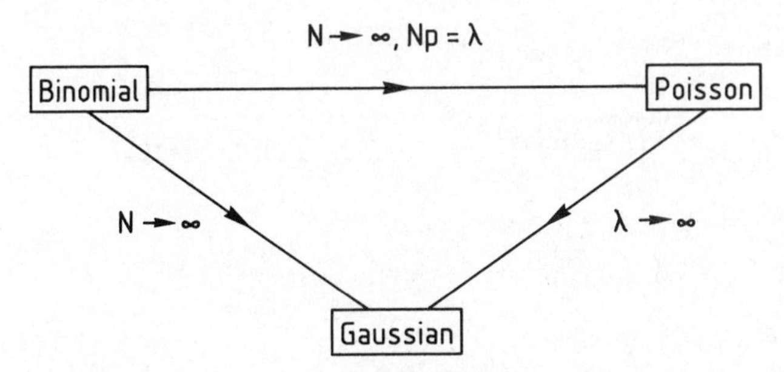

Fig. A4.2 The relationships among the binomial, Poisson and Gaussian distributions.

Detectors for charged nuclear particles provide an important application of the Poisson distribution. Many such devices rely on the particles creating photons as they pass through the detector, and these in turn release electrons in a photo-multiplier, which produces the required signal. For a large number of particles successively incident on a given counter, the numbers of such photo-electrons closely follow a Poisson distribution. If the counter is such that their mean number is low, there will be a significant probability that, because of fluctuations, in a given case the number of electrons released is zero, with the result that the counter gives no signal and the corresponding original particle is not detected. For example, for an average of 3.1 electrons, the probability of producing none is

$$P_o = e^{-3.1} = 4.5\%.$$

Hence the efficiency of the counter for detecting particles can be at best 95.5%.

Appendix 5

Student's t distribution

In Section 1.6, we considered the quantity

$$f = \frac{x - \mu}{\sigma} \tag{1.16}$$

where $x \pm \sigma$ is a measurement which is Gaussian distributed, and which we want to compare with some expected value μ. Then, in order to assess the significance of a non-zero value of f, we look up some tables of the area in the tails of the relevant probability distribution. This too will be Gaussian, if σ is the correct value of the experimental error.

Very often, instead of the true value of σ, we merely have an estimate u based on the observed spread of the measurements x_i. For a large number N of measurements, u should be close to σ, and the use of the Gaussian will be approximately correct. However, for only a few observations, u and σ can differ, and

$$f' = \frac{x - \mu}{u} \tag{A5.1}$$

follows a Student's t distribution, whose exact shape depends on N. Because of fluctuations in the estimate u in the denominator, Student's t distributions are wider than the Gaussian (see Fig. A5.1), and the effect is more pronounced at small N. Thus for a given value of f or f', the areas in the tails of Student's t distributions are larger than that for a Gaussian. A table of these fractional areas is given in Appendix 6; the number of degrees of freedom ν is $N - 1$ (one measurement is not sufficient to estimate a spread).

Both from Fig A5.1 and from Table A6.1, we can see that the Gaussian can be regarded as the limit of Student's t distributions as N becomes very large. For small N the difference can be important. Thus, for example, the probability of obtaining $|f| > 3$ is 0.3% for a Gaussian distribution, but the corresponding number is 3% for Student's t with $N = 6$, or 20% for $N = 2$.

In using eqn (A5.1), it is important to realise that if x is the mean of N measurements, then u should be an estimate of the error on that

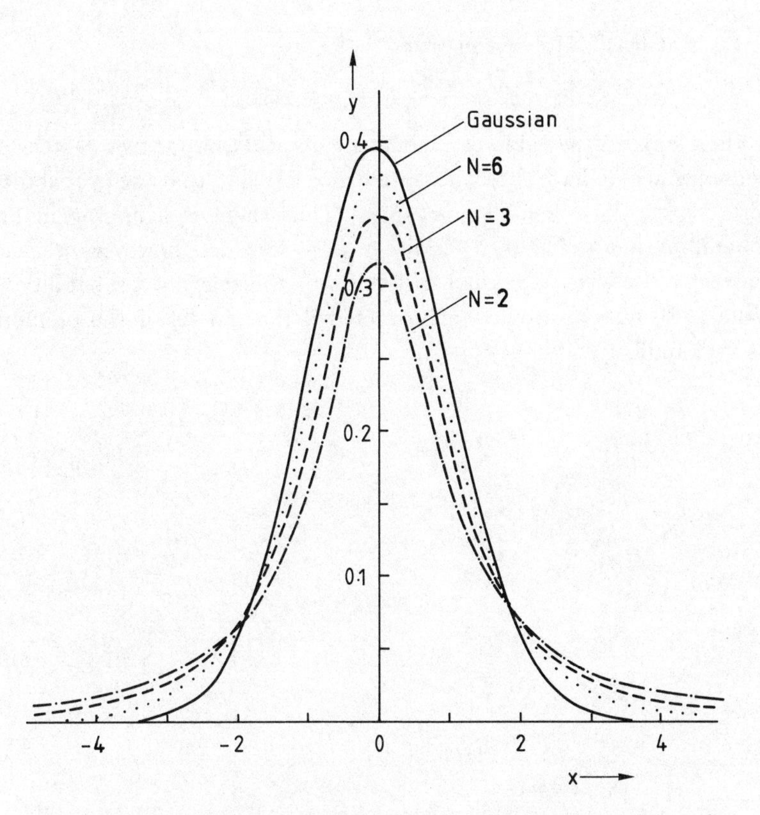

Fig. A5.1 Comparison of Student's t distributions for various values of the number of observations N, with the Gaussian distribution, which is the limit of the Student's distributions as N tends to infinity.

mean. This will as usual be a factor of $\sqrt{N}$ smaller than the spread of the x_i values, i.e.

$$u = s/\sqrt{N}$$

$$\text{where} \qquad s^2 = \frac{1}{N-1}\sum(x_i - \bar{x})^2. \qquad (1.3')$$

As a specific example, suppose someone says he has an IQ of 150. He is given three tests on which he scores 128, 143 and 113. What can we deduce about his claim? The average score is 128, and the estimated standard deviation s as deduced from the three measurements is 15, so the accuracy of the mean u is $15/\sqrt{3} = 9$. To test the suggestion that the IQ is at least 150, we construct

$$f' = \frac{128 - 150}{9} = -2.5.$$

We then look up the table of Student's t distribution for two degrees of freedom, since we have three measurements. We find that the probability of f' being -2.5 or smaller is $\sim$6%. Thus there is a non-negligible probability that the person's claim was correct. If, however, we had incorrectly used the Gaussian distribution, the relevant probability is around $\frac{1}{2}$%, and we would have concluded that an IQ of 150 or more was very unlikely.

Appendix 6

Statistical tables

This appendix contains tables relevant to the Gaussian, Student's t and χ^2 distributions.

Table A6.1 gives the fractional area f in one tail of the Student's t and of the Gaussian distributions, i.e.

$$f(r) = \int_r^\infty P(x)dx \tag{A6.1}$$

where $P(x)$ is the relevant normalised distribution, and we are interested in the fractional area beyond a value r of the variable x. For the Gaussian distribution,

$$P(x) = \frac{1}{\sqrt{2\pi}} \exp(-x^2/2) \tag{A6.2}$$

where for a measurement $m \pm \sigma$ when the true value is m_o

$$x = \frac{m - m_o}{\sigma}, \tag{A6.3}$$

i.e. it is the number of standard deviations by which m exceeds m_0. For Student's t,

$$x = \frac{m - m_0}{u} \tag{A6.4}$$

where u is the estimated standard deviation of m, based on the spread of N observations. These distributions depend on the number of degrees of freedom ν, which is $N - 1$.

As an example of using this table, we can look up the probability of being at least 2.0 standard deviations above the correct value. It is 2.3% for the Gaussian distribution, while for Student's t it is 3.7% for $\nu = 10$, or 15% for $\nu = 1$.

The Gaussian and Student's t distributions are all symmetric about $x = 0$. Thus the areas for $x > r$ are identical with those for $x < -r$. Similarly if we are interested in the fractional area in both tails of the distribution (i.e. the area for $|x| > r$), the numbers in the table should be doubled. Thus, according to the Gaussian distribution, the probability of deviating from the correct value by at least 2 standard deviations is 4.6%.

Some of the information in Table A6.1 for the Gaussian distribution is displayed in Fig. 1.7.

Student's t distributions have the property that, as the number of degrees of freedom increases, they tend to the Gaussian distribution; the numbers in Table A6.1 are consistent with this.

For further information, see Section 1.5 for the Gaussian distribution, and Appendix 5 for Student's t.

Table A6.2 is the inverse of Table A6.1, in that it gives the values of r corresponding to a specified fractional area $f(r)$ in the upper tail of the distributions. As in the previous table, the entries correspond to Student's t with various numbers of degrees of freedom ν, and to the Gaussian distribution. For example, there is only a 1% probability of being more than 2.3 standard deviations above the correct value for a Gaussian distribution, but for Student's t the corresponding number is 7.0 standard deviations for $\nu = 2$, or 2.5 for $\nu = 20$.

Finally Table A6.3 gives the fractional area in the upper tail of the χ^2 distribution, for ν degrees of freedom. (The χ^2 distribution exists only for positive values of the variable, and hence is not symmetric about zero (see Fig. 2.6).) This table corresponds in principle to Table A6.2 for the Gaussian and Student's t distributions. Thus there is a 5% probability of χ^2 exceeding 3.8 if $\nu = 1$, or 18.3 if $\nu = 10$. Some of the information in this table is displayed graphically in Fig. 2.7.

The use of χ^2 distributions is described in Section 2.8.

Table A6.1 *Fractional areas in tail of Student's t and Gaussian distributions*

r	ν								Gaussian
	1	2	3	4	5	10	15	20	
0.0	0.5000	0.5000	0.5000	0.5000	0.5000	0.5000	0.5000	0.5000	0.5000
0.1	0.4683	0.4647	0.4633	0.4626	0.4621	0.4612	0.4608	0.4607	0.4602
0.2	0.4372	0.4300	0.4271	0.4256	0.4247	0.4227	0.4221	0.4218	0.4207
0.3	0.4072	0.3962	0.3919	0.3896	0.3881	0.3852	0.3841	0.3836	0.3821
0.4	0.3789	0.3639	0.3580	0.3548	0.3528	0.3488	0.3474	0.3467	0.3446
0.5	0.3524	0.3333	0.3257	0.3217	0.3191	0.3139	0.3122	0.3113	0.3085
0.6	0.3280	0.3047	0.2954	0.2904	0.2873	0.2809	0.2787	0.2776	0.2743
0.7	0.3056	0.2782	0.2672	0.2613	0.2576	0.2499	0.2473	0.2460	0.2420
0.8	0.2852	0.2538	0.2411	0.2343	0.2300	0.2212	0.2181	0.2166	0.2119
0.9	0.2667	0.2316	0.2172	0.2095	0.2047	0.1946	0.1912	0.1894	0.1841
1.0	0.2500	0.2113	0.1955	0.1870	0.1816	0.1704	0.1666	0.1646	0.1587
1.5	0.1872	0.1362	0.1153	0.1040	0.0970	0.0823	0.0772	0.0746	0.0668
2.0	0.1476	0.0918	0.0697	0.0581	0.0510	0.0367	0.0320	0.0296	0.0228
2.5	0.1211	0.0648	0.0439	0.0334	0.0272	0.0157	0.0123	0.0106	0.0062
3.0	0.1024	0.0477	0.0288	0.0200	0.0150	0.0067	0.0045	0.0035	0.0013

Table A6.2 *Values of r for specified fractional areas in tail of Student's t and Gaussian distributions*

Area	ν								Gaussian
	1	2	3	4	5	10	15	20	
0.500	0.00	0.00	0.00	0.00	0.00	0.00	0.00	0.00	0.00
0.400	0.32	0.29	0.28	0.27	0.27	0.26	0.26	0.26	0.25
0.300	0.73	0.62	0.58	0.57	0.56	0.54	0.54	0.53	0.52
0.200	1.38	1.06	0.98	0.94	0.92	0.88	0.87	0.86	0.84
0.100	3.08	1.89	1.64	1.53	1.48	1.37	1.34	1.33	1.28
0.050	6.31	2.92	2.35	2.13	2.02	1.81	1.75	1.72	1.64
0.020	15.89	4.85	3.48	3.00	2.76	2.36	2.25	2.20	2.05
0.010	31.82	6.96	4.54	3.75	3.36	2.76	2.60	2.53	2.33
0.005	63.66	9.92	5.84	4.60	4.03	3.17	2.95	2.85	2.58
0.002	159.15	15.76	8.05	5.95	5.03	3.72	3.39	3.25	2.88
0.001	318.31	22.33	10.21	7.17	5.89	4.14	3.73	3.55	3.09

Table A6.3 *Values of chi-squared for specified fractional area in tail*

Area	ν									
	1	2	3	4	5	10	15	20	30	50
0.500	0.5	1.4	2.4	3.4	4.4	9.3	14.3	19.3	29.3	49.3
0.400	0.7	1.8	2.9	4.0	5.1	10.5	15.7	21.0	31.3	51.9
0.300	1.1	2.4	3.7	4.9	6.1	11.8	17.3	22.8	33.5	54.7
0.200	1.6	3.2	4.6	6.0	7.3	13.4	19.3	25.0	36.3	58.2
0.100	2.7	4.6	6.3	7.8	9.2	16.0	22.3	28.4	40.3	63.2
0.050	3.8	6.0	7.8	9.5	11.1	18.3	25.0	31.4	43.8	67.5
0.020	5.4	7.8	9.8	11.7	13.4	21.2	28.3	35.0	48.0	72.6
0.010	6.6	9.2	11.3	13.3	15.1	23.2	30.6	37.6	50.9	76.2
0.005	7.9	10.6	12.8	14.9	16.7	25.2	32.8	40.0	53.7	79.5
0.002	9.5	12.4	14.8	16.9	18.9	27.7	35.6	43.1	57.2	83.7
0.001	10.8	13.8	16.3	18.5	20.5	29.6	37.7	45.3	59.7	86.7

Appendix 7

Random numbers

Random numbers have an amazing variety of uses. Apart from the obvious one of specifying a random sample from a large population, they can be used for integrating complicated functions of one or more variables; for obtaining a value of π; or for simulating experimental errors. Further details can be obtained, for example, from Chapter 6 of the book *Statistics for Nuclear and Particle Physicists* by Lyons (Cambridge University Press 1986, ISBN 0 521 37934 2).

If you need a lot of random numbers, you can obtain them from a computer. For a smaller problem, such as that of Section 1.11.3, you can push the random number button on your calculator as many times as necessary, or use tables of random numbers. Those presented in table A7.1 were obtained as part of a sequence generated by a computer using an algorithm from the NAG Program Library.

Table A7.1

0.1612	0.2958	0.6909	0.6115	0.6209	0.2115	0.8305	0.4981	0.2051	0.0961
0.9388	0.9098	0.8596	0.9718	0.8607	0.2037	0.2646	0.2433	0.0318	0.6032
0.1345	0.0613	0.6015	0.8669	0.1079	0.3868	0.5100	0.2300	0.3669	0.8062
0.4399	0.4189	0.0148	0.0839	0.3481	0.1504	0.2313	0.7315	0.3094	0.6477
0.5785	0.8716	0.7120	0.4407	0.9223	0.5709	0.4016	0.9984	0.5847	0.8591
0.2739	0.3634	0.3653	0.3662	0.2160	0.6132	0.9123	0.2700	0.6408	0.2777
0.4614	0.0746	0.0569	0.5369	0.9407	0.6663	0.3203	0.4714	0.9378	0.0939
0.7225	0.7417	0.3379	0.1548	0.1515	0.6845	0.5846	0.8167	0.6203	0.8992
0.5858	0.0458	0.3375	0.8827	0.5732	0.2053	0.4636	0.8826	0.4528	0.3100
0.7394	0.2657	0.7778	0.7119	0.2182	0.3669	0.2081	0.7224	0.1129	0.0213
0.1607	0.7693	0.7823	0.4690	0.0980	0.5232	0.2354	0.7245	0.0619	0.4281
0.2687	0.3131	0.9598	0.3495	0.6878	0.6149	0.3892	0.2673	0.5639	0.1968
0.3635	0.4561	0.2886	0.6546	0.0748	0.9597	0.4565	0.4768	0.9572	0.0572
0.2641	0.9809	0.6272	0.2024	0.0214	0.4982	0.4116	0.9654	0.1684	0.1389
0.1058	0.2687	0.6454	0.2773	0.2393	0.4145	0.8990	0.6407	0.4038	0.1769
0.1674	0.9635	0.6010	0.6373	0.0044	0.5888	0.3945	0.1869	0.7371	0.3836
0.8531	0.8826	0.5368	0.2815	0.4324	0.6244	0.3804	0.5873	0.9573	0.3387
0.2020	0.1540	0.0870	0.2957	0.6696	0.9316	0.9384	0.9548	0.3122	0.6757
0.0872	0.2289	0.6819	0.8522	0.3901	0.2894	0.4249	0.8229	0.5960	0.1728
0.4729	0.3719	0.6808	0.9598	0.4786	0.6727	0.4902	0.0690	0.0733	0.2018
0.1849	0.5163	0.2151	0.0455	0.7537	0.2308	0.5504	0.9189	0.7696	0.9371
0.7597	0.7810	0.8958	0.6855	0.5540	0.3979	0.2132	0.5613	0.0359	0.3133
0.0650	0.3599	0.1393	0.9767	0.5535	0.0836	0.4033	0.1739	0.8736	0.5886
0.2389	0.6141	0.9962	0.9841	0.2672	0.6515	0.5714	0.7308	0.2171	0.7166
0.4188	0.0184	0.7254	0.1707	0.8784	0.1920	0.3240	0.8704	0.2340	0.5252
0.4594	0.4000	0.3174	0.4188	0.9678	0.1968	0.0329	0.5148	0.4766	0.3688
0.1726	0.7926	0.5123	0.5828	0.6666	0.7764	0.9587	0.9865	0.1770	0.9626
0.3008	0.0304	0.9913	0.4069	0.9316	0.6386	0.7219	0.9991	0.9032	0.5700
0.1734	0.8411	0.8532	0.6322	0.8427	0.5899	0.8193	0.6322	0.0845	0.0942
0.2106	0.1246	0.6176	0.9549	0.2748	0.8363	0.8345	0.7632	0.0634	0.0716
0.1043	0.6091	0.0818	0.5048	0.8626	0.4412	0.9736	0.2473	0.7989	0.7992
0.1429	0.8357	0.2775	0.4506	0.2962	0.4123	0.1572	0.3925	0.9652	0.8624
0.4139	0.8913	0.2977	0.4445	0.3477	0.0635	0.9668	0.7081	0.0157	0.4258
0.5174	0.2191	0.5630	0.4612	0.4434	0.1300	0.2953	0.3056	0.2425	0.0524
0.4051	0.1129	0.3927	0.1770	0.6462	0.0003	0.5597	0.7619	0.6612	0.4184
0.1301	0.5088	0.3599	0.3718	0.3404	0.2868	0.8224	0.3076	0.4711	0.3992
0.7187	0.3842	0.4778	0.1361	0.6790	0.8731	0.4961	0.8505	0.8059	0.4265
0.0065	0.7556	0.0019	0.2830	0.3683	0.0651	0.8126	0.8206	0.2881	0.6356
0.5198	0.0498	0.7710	0.5466	0.1398	0.1900	0.6566	0.3079	0.0033	0.6263
0.2021	0.4637	0.8667	0.0641	0.7881	0.3363	0.2271	0.4838	0.1638	0.0049
0.8949	0.2172	0.1421	0.3084	0.9007	0.2637	0.2759	0.0453	0.0900	0.4084
0.2387	0.3347	0.4122	0.0677	0.1981	0.5610	0.8705	0.8362	0.9670	0.6125
0.5572	0.0639	0.1584	0.2443	0.8561	0.0152	0.8113	0.4750	0.5148	0.7349
0.6456	0.3518	0.3714	0.9678	0.6351	0.6115	0.4538	0.5110	0.7641	0.9565
0.8290	0.4911	0.8552	0.6976	0.6200	0.2409	0.7636	0.7636	0.4136	0.4964
0.3947	0.7198	0.9570	0.8057	0.8231	0.1039	0.2032	0.5203	0.2186	0.3447
0.6978	0.8359	0.2945	0.5625	0.1368	0.7958	0.5927	0.5800	0.4394	0.4124
0.9002	0.1687	0.3502	0.8701	0.0745	0.8677	0.5476	0.1129	0.9703	0.7115
0.7175	0.6064	0.8326	0.0770	0.9791	0.8901	0.4206	0.4272	0.3607	0.2552
0.9446	0.3984	0.4359	0.9568	0.0764	0.1840	0.9092	0.7041	0.1368	0.9702

Index